OXFORD WORLD'S CLASSICS

EDWIN A. ABBOTT

Flatland

A Romance of Many Dimensions

Edited with an Introduction and Notes by
ROSEMARY JANN

UNIVERSITY PRESS

OXFORD
UNIVERSITY PRESS

Great Clarendon Street, Oxford OX2 6DP

Oxford University Press is a department of the University of Oxford.
It furthers the University's objective of excellence in research, scholarship,
and education by publishing worldwide in

Oxford New York

Auckland Cape Town Dar es Salaam Hong Kong Karachi
Kuala Lumpur Madrid Melbourne Mexico City Nairobi
New Delhi Shanghai Taipei Toronto

With offices in

Argentina Austria Brazil Chile Czech Republic France Greece
Guatemala Hungary Italy Japan Poland Portugal Singapore
South Korea Switzerland Thailand Turkey Ukraine Vietnam

Oxford is a registered trade mark of Oxford University Press
in the UK and in certain other countries

Published in the United States
by Oxford University Press Inc., New York

First published as an Oxford World's Classics paperback 2006
Reissued 2008

British Library Cataloguing in Publication Data

Data available

Library of Congress Cataloging in Publication Data

Data available

ISBN 978-0-19-953750-1

10

Typeset in Ehrhardt
by RefineCatch Limited, Bungay, Suffolk
Printed in Great Britain by
Clays Ltd, St Ives plc

CONTENTS

ACKNOWLEDGEMENTS

I AM grateful to Stephen Crook at the New York Public Library and Shaun Hardy at the Geophysical Laboratory Library at the Carnegie Institution of Washington for their assistance in enabling me to view first editions of *Flatland*. I also appreciate the assistance of *Flatland* experts Tom Banchoff and Bill Lindgren in helping me locate references to Abbott's involvement with women's education, and Jonathan Smith's help in locating reviews of *Flatland*. The sections on 'Science, Imagination, and Belief' and 'Flatland and Nineteenth-Century Geometries' in the Introduction are based in part on my previously published articles 'Abbott's *Flatland*: Scientific Imagination and "Natural Christianity" ', *Victorian Studies*, 28 (1985), 473–90 and 'Christianity, Spiritualism, and the Fourth Dimension in Late Victorian England', *Victorian Newsletter*, 70 (Fall 1986), 24–8.

INTRODUCTION

SINCE *Flatland* was first published anonymously in late 1884, it has earned a unique position in the genre of science fiction and fantasy. Although not even acknowledged in the 1937 *Dictionary of National Biography* entry for its author, Edwin Abbott Abbott (1838–1926), it has become the best-known work of this late Victorian educator and theologian. *Flatland* continues to charm modern readers by opening our imaginations to the possibility of a fourth and higher dimensions, but in its own day, it also participated in wider controversies about science, religion, and the social order. Its gentle satire of the blind spots of Abbott's Victorian contemporaries has continuing relevance to our own intellectual short-sightedness, as it encourages readers to recognize and to question our assumptions about what is logical, natural, and real.

Abbott, a Cambridge University graduate and Anglican clergyman, was best known as headmaster of the City of London School, a position he held from 1865 to 1889. By all accounts a gifted teacher, he wrote various studies of grammar and rhetoric and supported the broadening of the classical curriculum that was conventional to English public schools at the time through the addition of English literature and science. After his retirement at the age of 50, he went on to write numerous theological works, staking out a 'Broad Church' or liberal position on the literal truth of biblical accounts. If *Flatland* is in one sense another of the many pedagogical exercises he produced over his long career, it also reveals habits of mind that shaped Abbott's forays into controversies over science and theology. Its explanation of how life would be experienced in the two-dimensional world of its main character, 'A Square', offers readers practical lessons in Euclidean geometry. The inability of the beings whom the Square encounters from Pointland (which has no dimensions) and Lineland (which has only one) to imagine a reality higher than their own prefigures the Square's own resistance when he is initiated into the mysteries of the third dimension by an emissary

Sphere. His experiences also establish models for analogical reasoning that could help Abbott's contemporaries imagine the possibility of unseen realities, just as they help the Square eventually to accept the reality of higher dimensions. The Sphere's own arrogance in similarly refusing to believe that his own world may not be the highest one possible further underscores Abbott's cautionary tale to his readers not to assume that their own perception establishes the limits of all possible intellectual and spiritual realms, a lesson that the chastened Square has accepted in the Preface to the revised edition.

There was considerable British interest in the idea of a fourth dimension of space in the 1870s and 1880s. The attempts of mathematician Charles Howard Hinton to imagine what perception would be like for creatures in one, two, and higher dimensions in his 1880 essay 'What is the Fourth Dimension?'[1] may well have offered the immediate inspiration for *Flatland*'s similar investigation of the subject. What lifts Abbott's work into the realm of classic literature, however, is its ability to infuse an exercise in mathematical speculation with whimsical wit and profound satiric purpose. Lewis Carroll (Charles Dodgson), a more conventional mathematician and clergyman than Abbott, also put his learning into fictional play with delightful results in *Alice's Adventures in Wonderland* (1865) and *Through the Looking Glass* (1871). The ludicrous pomposity of the kings of Lineland and Pointland suggests that Abbott shared Carroll's comic skill for deflating human folly. And Carroll's interest in mirror imaging and symmetry in *Looking Glass* and his speculation about the love life of linear creatures in *Dynamics of a Particle* (1865) indicate

[1] Hinton's 'What is the Fourth Dimension?' was first published in *Dublin University Magazine*, 96 os (Michaelmas 1880), 15–34 and reprinted in the *Cheltenham Ladies' College Magazine*, 8 (Sept. 1883), 31–52. It was later released as a pamphlet and eventually incorporated into Hinton's *Scientific Romances, First Series*. See Ian Stewart, 'Introduction', *The Annotated Flatland: A Romance of Many Dimensions* (Cambridge, Mass.: Perseus Publishing, 2002), pp. xix–xxiii, for a discussion of the case for Hinton's influence on *Flatland*. K. G. Valente proposes an alternative origin for *Flatland*, as Abbott's rebuttal of an 1877 article in the *City of London School Magazine* that advocated too direct a correspondence between higher dimensions and religious truth; 'Transgressions and Transcendence: *Flatland* as a Response to "A New Philosophy" ', *Nineteenth-Century Contexts*, 26/1 (2004), 61–77.

that he was aware of contemporary debates about the fourth dimension; indeed, at least one contemporary reviewer likened *Flatland* to *Looking Glass* on the basis of their common interest in 'transcendental geometry'.[2] However, Carroll's delight in sheer nonsense and absurdity blunts more specific satiric intentions, and he makes no attempt to provide a naturalistic explanation for the bizarre phenomena that Alice encounters. If anything, he employs references to 'transcendental geometry' in order to deflate its supporters.[3] *Flatland* belongs more properly to that genre of speculative satire that includes Jonathan Swift's *Gulliver's Travels* (1726) and Samuel Butler's *Erewhon* (1872). Such works were early prototypes of what science fiction critic Darko Suvin calls 'cognitive parables': they create an alternative world, treat it with verisimilitude, and use it analogically to challenge the standards of the authors' own societies.[4] The literature of adventure, especially involving travel to exotic lands, was enjoying a vogue during the imperialistic expansion of England in the nineteenth century, and several reviewers also linked *Flatland* to this genre.[5] In the cognitive parable, however, the purpose of confronting the exotic is to estrange the traveller from his own reality and to force him to recognize the contingency of his own values and assumptions. To make this confrontation more pointed, the conventions of the traveller's own society often appear in exaggerated or inverted form in the new world that he explores, or its logic is extrapolated to illogical or ridiculous extremes. Like Gulliver, the Square serves initially as an unreliable narrator whose blindness to the faults of his own world—for instance, his unreflecting assurance that the inhumane practices of Flatland's ruling classes are logical, natural, and even divinely ordained—ironically reveals his limitations to the reader. Only as a result of having to experience alternative mores does he come eventually to acknowledge

[2] 'Notes and News', *Science*, 5 (1885), 184.

[3] Carroll's defence of conventional geometry is made clear in his *Euclid and his Modern Rivals* (London: Macmillan, 1885).

[4] Darko Suvin, *Victorian Science Fiction in the UK: The Discourses of Knowledge and Power* (Boston: G. K. Hall, 1983), 26.

[5] *The Spectator* (1884), 1583; Robert Tucker, 'Flatland', *Nature*, 31 (1884), 76.

the illogic and inhumanity of views he considered natural before. Unfortunately, like Gulliver, the Square also ends up alienated from his own society and unable to communicate his insights to his peers. Abbott was more fortunate, in that most of his reviewers recognized *Flatland*'s social satire, although they remained divided about its spiritual messages and the legitimacy of the 'transcendental geometry' it endorsed. The ultimate target of this kind of parable is of course the readers, who by recognizing the narrator's blind spots should be led to question the inevitability and wisdom of their own conventional beliefs and practices—or as the Square puts it in his dedication, to develop 'that most rare and excellent Gift of Modesty' about their society's access to ultimate truth.

The Shape of Society in Flatland

As the Square is quoted as saying in the Preface (p. 10), not every detail in the social life of Flatland should be assumed to have a correspondence in Abbott's own society. However, there is much that is suggestive of Victorian conventions in this imaginary world. The attitudes that underpin the highly stratified class system of Flatland, in which rising status correlates strictly with increasing size and number of angles, reflect in various ways upon the acute class-consciousness of Abbott's Victorian contemporaries. Theirs was a society in which the smallest nuances of conduct, speech, and appearance were scrutinized for evidence of one's relative social standing. Flatlanders' anxieties about correctly identifying class identity by hearing (for instance, their fears that the working-class Isosceles triangles might successfully counterfeit the distinctive accent of the upper classes, p. 31) or about gauging the precise rank of polygons in the social hierarchy without resorting to the vulgar practice of feeling their angles (pp. 40, 59) provide a comic perspective on Victorian concerns about how to measure and to verify status, especially in the complicated hierarchy of the British peerage. The Flatland aristocracy, like their Victorian counterparts, benefited from a certain mystification of their claims to superiority (p. 59), by implying that these rested not on material measures like the exact number of their angles but on an incalculable essence of nobility. The elite's mastery of the finer points of 'Sight Recognition' at 'the illustrious University of Wentbridge' (Abbott's pun on Cambridge, p. 39) further strengthens their control over social power and social exclusion, as the Square ruefully notes on pp. 39–40.

Although many Victorians were as convinced as the Square that class differences correlated with absolute differences of character and ability, that did not mean that they considered these distinctions to be necessarily permanent. England had traditionally considered itself a society of 'removable inequalities', in the words of mid-century political commentator Walter Bagehot, one in which moral and material achievements could over time justify rising status. The history of the nineteenth century did in fact record the steady progress of the English middle classes in social and political power, as their growing importance in industry, commerce, and professional life translated into increased voting rights and access to the kinds of education and culture once open only to the gentry and aristocracy. It is significant, however, that the 'Law of Nature' (p. 21) in Flatland that allows each male child to gain one more side than his father and hence to advance with each generation toward the polygonal status of the nobility sanctions a form of social progress that is automatic only for the professional or gentlemanly squares and higher ranks. The working-class Isosceles triangles, if they increase the size of their angles (their social status) at all, usually do so only in half-degree increments per generation (p. 34). The Square is largely silent on how the equilateral or tradesman class develops into the square or gentlemanly one (although he mentions that his father is a triangle on p. 38). These limitations on triangular progress offer an ironic comment on the course of mid-Victorian class relations. Although the middle classes had been willing to make common cause with workers and tradesmen in agitating for the vote in the years preceding the First Reform Act (1832), once they had achieved it, they allied their interests with those of their social betters, as had indeed been the intent of the upper classes in selectively granting them the franchise in the first place. Like their Flatland counterparts, the Victorian middle classes only too willingly emphasized the continuum between their status and that of their betters while stressing their differences from the classes below. This included enforcing an important symbolic distinction between the lower-middle classes in trade (the equilateral triangles) and the upper-middle-class professionals and gentlemen, to whom Abbott assigns a different shape/class. This enforcement became all the more anxious as the extension of educational and cultural opportunities to the lower and lower middle classes after the Education Act of 1870 increasingly blurred this distinction. The lower-middle classes were equally invested in emphasizing their superiority to what they considered to be the less respectable classes below them. Conveniently for the Flatland status quo, any progress toward respectable equilateral status that a lower-class Isosceles could hope for was contingent upon behaviour that served the state (like success through military service) or that reinforced conformity and thus social stability, like hard work, frugality, and self-control (p. 22). The Flatland 'Law of Compensation' (p. 23), ensuring that as Isosceles gain in intelligence, their angles grow larger and make them less dangerous to their betters, does in a sense reflect the growing importance of respectability among the upwardly mobile artisan classes of Victorian England. However, in the Square's assurance that class differences are not just determined by natural law but are also divinely ordained (p. 23) we should recognize Abbott's satire on the readiness of his contemporaries to attribute biological and divine sanction to socially constructed (and highly self-interested) distinctions.

Moreover, notwithstanding the Square's sanguine confidence that such 'Laws of Compensation' guarantee safe and gradual progress, Flatland's upper classes, like their Victorian counterparts, remain persistently anxious about the insurrectionary potential of the lower classes. While British history records nothing like the 120 rebellions and 235 lesser outbreaks (p. 24) in Flatland's past, the first half of the nineteenth century was overshadowed by fear that the English working classes would rise in rebellion as the French had done in 1789. The radical egalitarianism of Flatland's 'Colour Revolt' suggests parallels to the French Revolution,[6] as does its collapse as participants are turned against one another by the manipulation of the polygons. Industrialization provoked strikes and numerous other conflicts between workers and owners during the first half of the century, many of them put down by violence. Disorderly agitations for the widening of the franchise during the Chartist movement of the 1840s and preceding the Second Reform Act in 1867 put middle-class anxieties about social order on edge, and the rise of socialism and anarchism in the 1880s kept them there. Abbott's own era was particularly concerned about the growing urban underclass, the impoverished 'residuum' that was feared as a source of crime and degeneracy. Many Victorians opposed efforts at charitable relief for fear that it would foster the unfit and endanger public safety. Abbott's rather different sympathies are hinted at by the controversy over a sermon he preached at Westminster Abbey, which was criticized by a church dignitary for 'inciting the poor against the rich'.[7] The Square's ready acquiescence in the Circles' cynical system for quashing insurrection by co-opting working-class leaders (p. 23), and his callous endorsement of practices like allowing the lower classes to starve (as in the case of the degraded Isosceles kept as models for practising sight recognition in schools, pp. 34–5) or to destroy one another as cost-efficient ways of keeping down the numbers of the potentially dangerous poor, afford Abbott the means of ironically exposing the limitations of a social logic shared by many of his contemporaries.

Justifying the Status Quo

The unreflecting respect for authority displayed in the Square's conviction that it was best for 'the interests of the Greater Number' of Flatlanders that irregular figures be 'painlessly and

[6] Suvin also sees elements of Roman history, Wat Tyler's rebellion, and nineteenth-century struggles for the vote in the Colour Revolt: Suvin, *Victorian Science Fiction in the UK*, 372.

[7] Quoted in A. E. Douglas-Smith, *The City of London School*, 2nd edn. (Oxford: Blackwell, 1965), 163.

mercifully consumed' (pp. 44–5) gestures more widely toward the powerful pressure to conform in Victorian society. In chapter 3 of *On Liberty*, John Stuart Mill's 1859 plea for toleration of difference, he lamented the tyranny of majority opinion, comparing the force of custom to the warping effect of a tiny shoe on the foot of a Chinese woman. The Square, on the other hand, argues that 'toleration of Irregularity is incompatible with the safety of the State', because without the complete predictability of status provided by regularity of configuration, social order would break down, and conventions like the size and shape of Flatland dwellings would have to be modified in order to accommodate irregular 'monsters' (p. 44). The tyranny of majority opinion is most apparent in the Circles' strict suppression of heretical views about higher dimensions, views that could undermine their claims to absolute superiority. In the end, of course, the Square falls victim to the same spirit of conformity that he had earlier endorsed; he cannot reveal his new insights to his sons for fear that their unquestioning loyalty to the Circles might overcome 'mere blind affection' (p. 112) to betray his seditious sentiments, and even his precocious grandson is too cowed by the Circles' authority to entertain the possibility of a third dimension that the Square now recognizes as real (pp. 112–13).

The Square's confidence that exterior form mirrors interior character also drew support from a range of supposedly scientific data that were used to validate contemporary social hierarchies during this period. Flatlanders' assumption that the size and number of an individual's angles correlate not just with class but also with moral and intellectual status has direct links to pseudo-sciences like physiognomy and phrenology, both of which enjoyed wide popular credibility during much of the nineteenth century. The former assumed that moral, emotional, and mental characteristics could be predicted from one's facial features, the latter from the size and shape of one's skull. Such thinking combined with the vogue of evolution (which was given dramatic support in Charles Darwin's 1859 *The Origin of Species*) to reinforce racial and class stereotyping. Thus scientists purported to demonstrate the 'natural' inferiority of savage races from their supposedly more animalistic facial angle and features, and criminologists like Cesare Lombroso used composite photographs to argue that criminals constituted a distinct biological type identifiable by its regression to more primitive physical traits. Appearing to provide biological sanction for the status quo, such pseudosciences in actuality helped to construct and to rationalize the hierarchies that they purported merely to describe.

Flatland attitudes toward irregularity are also implicated in long-standing debates about whether nature or nurture had more influence over individual character. Although the Square asserts that he never met an Irregular who 'was not also what Nature evidently intended him to be'—a hypocrite, misanthropist, and trouble maker—Abbott has him acknowledge the counterargument as well: that the perverted characters of Irregulars might result from the ill treatment they encounter (pp. 43–4). The nature/nurture debate gained added salience in the second half of the nineteenth century under the influence of evolutionary theory and the development of eugenics. As noted earlier, notwithstanding their belief in the quasi-biological basis of class differences, Victorians also celebrated the importance of effort, will, and hard work in improving one's standing in society. Their belief that inequalities were removable was enshrined in the Victorian best-seller *Self-Help* (1859) by Samuel Smiles, a work that rooted individual progress in striving and self-discipline. The ability of Isosceles gradually to increase their angles as a result of 'diligent and skilful labours' (p. 22) credits such attitudes. In evolutionary terms, this confidence in the ability of effort to lead to self-improvement was enshrined in the teachings of Jean-Baptiste Lamarck, who argued (erroneously) that traits acquired by parental effort could be passed on to offspring. The automatic social advance from generation to generation enjoyed by Flatland's higher classes embodies the kind of progressive interpretation that was popularly attributed to evolution, notwithstanding Charles Darwin's emphasis on struggle and adaptation and his conviction that a creature's innate fitness could not be altered by striving.

As the century wore on, the more pessimistic implications of

Darwin's theories gained strength, and confidence in Lamarckian progress waned. In *The Descent of Man and Selection in Relation to Sex*, his 1872 sequel to *The Origin of Species*, Darwin expressed concern that civilized societies blunted the progress of the race by keeping the unfit alive, although he also felt that this protection of the weak could not be suppressed without destroying 'the noblest part' of human nature,[8] a view the humbled Square comes in effect to share after his encounter with the Sphere. It was left to Darwin's cousin, Sir Francis Galton, to draw out the full implications of inheritance for the social order, starting in works like *Hereditary Genius* (1869). Rejecting the idea that environment or striving played any significant role in one's achievement, Galton argued that social distinction was determined by innate and inheritable traits. It was the responsibility of society to cultivate excellence by promoting the mating and reproduction of the most fit (and, by extension, by suppressing the breeding of the unfit—the mentally or physically defective, but often, by extension, the poor and unemployed, whose failure to succeed economically was taken as proof of their inherent inferiority). Galton coined the term 'eugenic' to describe this process in 1883.[9] As is also the case among Flatlanders, for whom social status is tacitly equivalent to the amount of intelligence indicated by a figure's angles, Galton tended to elide social, moral, and intellectual distinction, assuming that the achievement of 'eminence' in one's society was the direct result of inborn intellectual ability. Like Galton, the Square approves of the suppression of 'ancient heresies' that 'conduct depends upon will, effort, training, encouragement, praise, or anything else but Configuration' (p. 61); however, he is uncomfortable with the kinds of ethical dilemmas that result from such a strongly deterministic model and ultimately endorses the continuing importance of moral suasion on behaviour (p. 62). We can hear the voice of the Victorian

[8] Charles Darwin, *The Origin of Species by Means of Natural Selection, or, The Preservation of Favored Races in the Struggle for Life and The Descent of Man and Selection in Relation to Sex* (New York: Modern Library, 1936), 502.

[9] Francis Galton, *Inquiries into Human Faculty and Its Development* (London: Macmillan, 1883), 24–5.

eugenicist in the Square's anxieties about the 'extraordinary fecundity of the Criminal and Vagabond Classes' (p. 34) and in his support for controlling it through selective breeding by the Circles or by the 'providential' self-elimination of this 'redundant population' and of the potential for revolution that such dangerous classes were considered to harbour (p. 28). The counterpart to Victorian fears about the proliferation of the unfit 'residuum' was concern that the talented upper classes were not reproducing quickly enough, a fear confirmed in Flatland by the decreasing fertility of the Circular class as its members gain additional sides (pp. 59–60). The Neo-Therapeutic Gymnasium, where Polygonal parents have their children's frames reset in the hope of accelerating their ascent to circularity (p. 60), foreshadows the grim results of selective breeding inspired by the eugenic movement in later years.

Satire often proves to be a slippery weapon. Apparently some of the readers of *Flatland*'s first edition failed to grasp the irony that Abbott intended to aim at the Square for his whole-hearted endorsement of the ruthless and dismissive attitudes of the aristocracy toward the lower classes (an error also made by many readers of Jonathan Swift's satiric essay 'A Modest Proposal'). In his 'Preface to the Second and Revised Edition' of *Flatland*, Abbott corrects this impression by having the Square undergo a change of heart after his long imprisonment and recognize the infecundity of the Circles as a judgement by Nature against their world view (p. 10).

Female Flatlanders

The 'Preface' also disavows the Square's slighting views of women in Flatland, views that some contemporary readers similarly attributed to Abbott himself (see for instance the review in the journal *Nature*, which speculates that the Square must have 'suffered a disappointment at the hands of a lady').[10] In a world where geometrical configuration is everything, female Flatlanders

[10] Tucker, 'Flatland', 77.

are lower in intelligence than even the sharpest-angled Isosceles, being straight lines, even among the aristocracy. Their irrationality makes their sharp ends profoundly dangerous to their husbands and children and thus justifies restrictions on their movement and behaviour. Just as science had been pressed into use to prove the 'natural' inferiority of criminals and savages to upper-class Europeans in the nineteenth century, so too was it used to demonstrate that woman's inferiority to man was the product of her inherent biological nature, rather than being, as liberals like J. S. Mill argued, the result of culturally restricted roles. The complete stasis in female mental development that the Square notes—' "Once a Woman, always a Woman" is a Decree of Nature; and the very Laws of Evolution seem suspended in her disfavour' (p. 29)—found support in Darwin's thinking about the female's evolutionary retardation in *The Descent of Man*.[11] Evolutionary biologist George Romanes codified widely held views about the 'Mental Differences between Men and Women' in his 1887 essay of the same title. According to Darwinian theory, during the development of the human race, male competition for females naturally honed male strength and intelligence while depriving females of most of the benefits of evolution. The supposedly smaller size of women's heads (what Romanes referred to as the 'missing five ounces of the female brain'[12]) was offered as empirical proof of their weaker mental power and evidence that their mental disabilities could not be quickly corrected. In these ways conventional stereotypes about women—that they are controlled by caprice and emotion (as the Square confirms in saying that his wife possessed 'the usual hastiness and unreasoning jealousy of her Sex' (p. 82), or that they are naturally deficient in will, originality, and judgement—were given scientific sanction.

Ideally, the Victorian doctrine of 'separate spheres' held that middle-class men and women had distinct but complementary

[11] Darwin, *The Origin of Species*, 873–4.

[12] George Romanes, 'Mental Differences between Men and Women', in Dale Spender (ed.), *The Education Papers: Women's Quest for Equality in Britain 1850–1912* (London: Routledge, 2001), 23. Romanes's essay originally appeared in *Nineteenth Century*, 21 (May 1887), 654–72.

functions: men were charged with business and governance, as befitted their greater intellectual and physical strength, while women's natural propensity for affection, sympathy, altruism, and piety destined them to be the managers of family and home. In practice, however, this belief system characterized women's attempts to gain education, voting rights, and employment outside the home as unnatural and dangerous to their health. Although education for middle-class girls was not completely ruled out as is the case in Flatland, where the Chief Circle long ago had decreed that 'since Women are deficient in Reason but abundant in Emotion, they ought no longer to be treated as rational, nor receive any mental education' (p. 64), their instruction heavily emphasized finishing school deportment and 'accomplishments' like dancing, music, and needlework. Women were excluded from Oxford and Cambridge until the 1870s and could not earn degrees there until the 1920s. As an educator, Abbott strongly supported attempts to expand educational opportunities for women.[13] As a progressive thinker, he targeted the hypocrisy of separate spheres ideology, which kept middle-class women on a pedestal at the cost of denying them full rationality. As a theologian, Abbott further condemned the hypocrisy that led men to indulge women with talk of 'love', 'duty', and 'right' at home while replacing these with 'the anticipation of benefits', 'necessity', and 'fitness' in the competitive and self-interested marketplace (p. 64). The chastened Square, who in the first half of *Flatland* considers this kind of double-speak merely a nuisance to men (p. 65), eventually accepts (p. 9) the Sphere's teaching that qualities like love and mercy, condemned as feminine in Flatland, are more important than intellect in assessing human merits (pp. 97–8).

[13] Abbott was a guest on the platform at the meeting that inaugurated the Girls' Public Day School Company, which Maria Grey founded to extend educational opportunities for girls in 1872. He further assisted Grey in her efforts to develop better training for teachers. For information on his support for women's education, see Edward W. Ellsworth, *Liberators of the Female Mind: The Shirreff Sisters, Educational Reform, and the Women's Movement* (Westport, Conn.: Greenwood Press, 1979), 184–5, 214, 220.

Science, Imagination, and Belief

The fact that Abbott labels the male vocabulary of calculation and self-interest part of the 'idiom of Science' (p. 65) points to the more serious intellectual goals of this deceptively light-hearted work. *Flatland* participated in a debate about the limits of human knowledge that embraced science, mathematics, and religion in the second half of the nineteenth century. A range of scientific developments challenged Christian orthodoxy during the Victorian period. Fossil discoveries raised questions about the accuracy of biblical chronology. Evolution challenged belief in the special creation of humans and animals by a benevolent divinity. The historical and comparative study of myth undermined the credibility of biblical miracles and the Christianity they supported. This challenge to orthodox faith was felt by scientists as well as laypeople. Some scientific idealists like William Whewell kept God in the universe by asserting that scientific concepts provided access to a transcendental and divinely ordered truth. In response, materialists like Thomas Henry Huxley questioned the human ability to obtain truth about matters beyond the reach of experience. He also coined the term 'agnostic' to label the unknowability of the divine and to indicate that he considered theological truth beyond the reach of any meaningful scientific proof. Some men of letters like Matthew Arnold acknowledged the fallibility of biblical accounts but tried to preserve their spiritual value as literature. Abbott felt bound to acknowledge the norms of scientific and historical truth current in his day but was unwilling to accept different standards of truth for the rational and the spiritual. The key to his solution of this dilemma lay in the imagination, the same kind of imagination that allows the Square ultimately to escape the limits of his own perceptions and to recognize the possibility of higher realities.

Imagination played a pivotal if slippery role in various Victorian debates about the limits and possibilities of human understanding. The hallmark of English science had traditionally been empirical investigation and inductive reasoning, as championed by Sir Francis Bacon in the sixteenth century and by Sir Isaac

Newton in the eighteenth. Newton's famous statement in his *Principia Mathematica*, 'Hypotheses non fingo' ('I make no hypotheses'), implied that true science was a matter of careful observation and measurement, not of fanciful theories. In fact, both Bacon and Newton had employed hypotheses, and scientific discovery is largely impossible without doing so. Although during the course of the nineteenth century, various philosophers of science moved away from simplistic conceptions of Baconian induction to acknowledge the role of imagination in scientific understanding, much of the public remained suspicious of theorizing, especially when they saw how alarming the results could be in Darwin's theory of evolution. Scientific popularizer John Tyndall was replying to this kind of scepticism in 1868 when he looked forward to continued intellectual progress in terms that anticipated some of Abbott's own arguments. Tyndall argued that science and myth relied equally on imaginative leaps beyond immediate experience and counselled intellectual humility in a distinctly Flatlandish metaphor: just as two-thirds of the sun's rays were now invisible to the naked eye, there might exist vast reaches of knowledge requiring only the development of the 'proper intellectual organs' to perceive them.[14]

Abbott, a scholarly specialist in the work of Sir Francis Bacon, was clearly aware of such contemporary debates but sought to respond to them in ways that stressed the compatibility of scientific and spiritual reasoning. As an Anglican priest, he was considered a proponent of what were called 'Broad Church' sentiments, which included applying to biblical texts the same kinds of historical and linguistic principles used to analyse secular documents. He believed Christian conduct to be more important than literalist interpretations of the Bible; indeed, he felt that insisting on the literal truth of miracles that clearly violated scientific law was more likely to weaken faith than to strengthen it. In theological works written around the time of *Flatland*, he elaborated an interpretation of imagination in human history that

[14] John Tyndall, 'Scientific Materialism', in *Fragments of Science* (New York: Appleton, 1897), ii. 89.

had the effect of putting religious belief on the same footing as scientific concepts. In *The Kernel and the Husk* (1886), he argued that our belief in the continuing uniformity of nature was simply a leap of the imagination, tested against experience. We could confirm the functioning of what we called 'force' or 'cause' but could prove neither to be real entities. John Tyndall, like his fellow materialists, might concede the same, but he clearly expected such concepts eventually to yield to advancing human intellect, and like Huxley he considered ultimate spiritual realities to be by definition unknowable. Abbott stressed that faith worked in the same way as science: just as we could believe in scientific concepts simply because they 'worked', not because we could prove them, we could as confidently believe in religious concepts so long as these 'worked' to make us better people. In *Through Nature to Christ, or, The Ascent of Worship Through Illusion to the Truth* (1877), Abbott argued that rather than revealing nature's truths directly, God had during the course of human history provided misleading 'illusions' in order to develop man's truth-seeking faculties. Man's relationship to a nature whose full truth he could but glimpse nurtured a belief in more than could be logically proved by material evidence. Humanity's struggle to interpret natural phenomena in primitive times had required constant leaps of the imagination that providentially prepared them for their eventual reception of the truths of Christianity. By thus positing imagination as the basis of all knowledge, Abbott sanctioned a religion independent of material proof: there was no need to require violations of physical laws—miracles or even Christ's resurrection—in order to believe in the higher, spiritual truths of Christianity. Just as God revealed nature's mysteries only in glimpses, his manifestations of himself in human history were never more than 'refractions' of an immaterial reality, illusions subject to successive reinterpretations over time, each of which got closer to the truth. Biblical accounts of miracles thus had the same status as early scientific theories: both represented the attempts of earlier cultures to explain illusions in terms they could understand. Such interpretations allowed Abbott to reject the limits that scientific materialists would place on our access to religious truths but also to oppose what he considered a blind and anti-intellectual surrender to authority such as that represented by John Henry Newman's orthodoxy about miracles and church dogma.[15] For Abbott, the glory of Christianity was that the constant challenge of illusion had kept faith from degenerating into a new enslavement to law.

Seen in the light of such views, *Flatland* can also be read as an allegory aimed at correcting the arrogance of both the materialist intellect and dogmatic faith and at demonstrating the progressive force of imagination. The insistence that 'Feeling is believing' (p. 82) links females and the lower classes in Flatland to a narrow religious fundamentalism, incapable of venturing beyond literal interpretations. Sight recognition, taught at the university, represents an intellectual advance by embodying a process of induction and inference that Abbott considered necessary to a higher understanding of the truths concealed by appearances. All beings in the tale stand condemned for their failures of imagination, however, and for their arrogance in assuming that what they can perceive constitutes the whole of reality. The fatuous solipsism of the kings of Pointland and Lineland, who assume that their kingdoms constitute all of space, is duplicated in the Square, the Circles, and the Sphere as well. The Square ridicules both kings for their ignorance of 'reality', only to find the same arguments turned against him by the Sphere. Next it is the Sphere's turn to dismiss as 'utterly inconceivable' (p. 103) the reality of dimensions beyond his perception. The Square has clearly been brainwashed by the Circles into believing that their equation of dimensionality with social worth is ordained by natural law, if not actually constituting a caste system of 'divine origin' (p. 23). It is only after he literally experiences the reality of higher dimensions that he can understand the fallibility of the Circles' fetishizing of 'omnividence' (p. 97) and see their ultimate defeat through infecundity as providential. The conclusion he draws about his experiences is similar to Abbott's view of the history of religious belief:

[15] Abbott's attacks on Newman's theological positions shortly after his death engaged him in a highly contentious public controversy with Newman's defenders in the early 1890s. See the Chronology for specific titles.

'herein . . . I see a fulfilment of the great Law of all worlds, that while the wisdom of Man thinks it is working one thing, the wisdom of Nature constrains it to work another, and quite a different and far better, thing' (p. 10). The Square thus undergoes his own personal journey through illusion to truth and realizes that the leap of faith (pp. 7–8) necessary to interpret his own world lends equal validity to higher realms, even if he can experience them only in 'Thoughtland' (p. 106).

Flatland and Nineteenth-Century Geometries

The challenge of understanding higher dimensional geometries that *Flatland* investigates occupied a central position in Victorian debates about the accessibility of absolute truth. As Joan Richards has demonstrated, geometry served as the 'queen of the sciences'[16] in the nineteenth century and was central to debates over the nature of human knowledge. Euclid's famous mathematical treatise *Elements* had traditionally formed the backbone of education in England, a tradition in which mathematics was considered not a specialized branch of knowledge, but a model for all advanced reasoning. The self-evidence of Euclidean geometrical axioms and their predictive accuracy endowed mathematics with a necessary truth that modelled the certainty of God's existence. Nineteenth-century idealists like William Whewell used geometry to buttress their claims that scientific concepts gave us access to a higher reality beyond appearances. The opposing materialist or empiricist view treated such concepts simply as convenient mental constructs describing or summing up previous observation, yielding no access to transcendental truth. In this view, it was at least possible that the sun might not rise tomorrow, no matter how unlikely. But in Euclidean terms, such violations of law were impossible, like a triangle whose angles totalled more than 180 degrees. Empiricists wishing to treat geometry as simply a logically consistent formal system were thwarted by the

[16] Joan Richards, *Mathematical Visions: The Pursuit of Geometry in Victorian England* (Boston: Harcourt Brace, 1988), 2.

privileged position of Euclidean axioms as seemingly necessary truths.

This situation changed with the introduction of non-Euclidean geometries in the early nineteenth century by European mathematicians like Carl Friedrich Gauss, Nickolai Lobachevsky, and János Bólyai. In his later popularization of such alternative models, German mathematician Hermann von Helmholtz was explicitly trying to demonstrate that Euclidean geometry was not the only way of explaining the behaviour of space, thus challenging its claim to absolute truth. In an 1870 article on 'The Axioms of Geometry' published in the English periodical *The Academy*, Helmholtz imagined how space would be perceived differently by two-dimensional creatures living on a plane like Flatland and by those sliding along the surface of a sphere. Although at close quarters the Euclidean geometry of the plane dwellers might also hold true for the sphere dwellers, as the latter gained wider experience of their world they would encounter straight lines that intersected at more than one point (as they crossed at the poles of the sphere) and triangles mapped onto the round surface whose angles would add up to more than 180 degrees. In their spherical world, Euclidean axioms held true only in small, localized spaces and could claim no necessary or absolute truth.

Although, for mathematicians like Helmholtz, the possibility of higher dimensions challenged Euclidean claims to reveal a necessary and transcendent truth and thus buttressed empiricist arguments, other scientists seized upon the fourth dimension as a means of affirming the reality of the spiritual and the supernatural. In works like *The Unseen Universe* (1875) and *Paradoxical Philosophy* (1878) Peter Guthrie Tait and Balfour Stewart attempted to justify Christian belief in God and immortality by imagining a fundamental continuity between our visible universe and a spiritual one in the fourth dimension. Physicist James Clerk Maxwell imagined his soul as a trefoil knot that (according to the theorizing of German mathematician Felix Klein) could be untied only in the fourth dimension. The most notorious attempt to spiritualize higher dimensions involved the German astronomer

Friedrich Zöllner, who became convinced of the reality of the fourth dimension after being duped by the fraudulent tricks of the notorious spiritualist medium Henry Slade, like causing knots to appear in a closed loop of string. In *Transcendental Physics*, which appeared in English translation in 1880, Zöllner insisted that the fourth dimension could explain not only spiritualist manifestations but Christian miracles as well.

Abbott's own theological objectives dictated a complicated response to contemporary debates about Euclidean and non-Euclidean space and the possibilities of higher dimensions. As a member of the Association for the Improvement of Geometrical Teaching, he joined forces with other educators in trying to devise pedagogical alternatives to Euclid's *Elements*. The experiences that he has the Square undergo in *Flatland* certainly challenged the privileged status of three-dimensional geometry. And in later works like *The Spirit on the Waters*, Abbott argued explicitly that whatever the predictive power of mathematics, it did not allow us direct access to the divine.[17] Yet Abbott's own purposes were also served by some of the philosophical positions staked out by Euclid's supporters. Those who wanted to defend the transcendental truth of Euclidean axioms stressed the distinction between being able logically to work out an understanding of what alternative geometries might be like and being able actually to conceive or visualize their reality. For such idealists, a formal working out of the properties of non-Euclidean geometries did not challenge the privileged status of Euclidean space, which was still the only one conceivable given the perceptual limits imposed by our three-dimensional brains. Abbott endorsed this distinction in *The Kernel and the Husk*, noting that we cannot 'conceive of space of Four Dimensions . . . although we can perhaps describe what some of its phenomena would be if it existed'.[18] The inability of the kings of Pointland and Lineland to imagine worlds beyond their own are failures of 'conceiving' in this sense. The Sphere teaches the Square to reason out by analogy what

[17] Edwin Abbott Abbott, *The Spirit on the Waters: The Evolution of the Divine from the Human* (London: Macmillan, 1897), 32.

[18] Edwin Abbott Abbott, *The Kernel and the Husk* (London: Macmillan, 1886), 259.

spatial reality would feel like in the third dimension, but it is only when he physically lifts the Square into the next dimension that the Flatlander can fully conceive this higher dimension. Once returned to Flatland, the Square experiences increasing difficulty in trying to reconstruct its reality in his mind. The fact that he must finally depend upon faith (pp. 7–8) to affirm its existence suggests that his analogical understanding of how Spaceland must operate does not give it a conceivable reality.

And yet this leap of faith does not invalidate the Square's experiences, since for Abbott, geometrical truth depends on the same acts of imagination as do other forms of human understanding and indeed formed a model for it. As his hypothetical geometer argues in *The Kernel and the Husk*, no 'chalkland' triangle was exactly equilateral, no chalkland point literally of one dimension. Although this mathematician had never seen a perfect circle, to him it was 'as real as a beefsteak and a pint of porter' in so far as it 'worked' to predict correctly the behaviour of reality. He accepted its existence 'by Faith' and believed that God 'intended us to study this and other immaterial realities that our minds might approximate to His' (p. 32). Thus for Abbott, geometry was a model for advanced reasoning not because it offered direct access to truth, but because it depended on the same forms of imagination that religion did. The Square's understanding of higher dimensions is affirmed in so far as it duplicates the struggle through illusion to a grasp of higher reality that Abbott saw as the providential course of all human cognition.

It might seem likely that Abbott would make common cause with those who wished to use the fourth dimension to explain supernatural phenomena. Several reviews of *Flatland* linked it to spiritualism, and at least one later work, Alfred Taylor Schofield's *Another World, or, The Fourth Dimension*, explicitly relied on *Flatland*'s authority to buttress its own argument about the reality of the spirit world.[19] The Sphere's grudging admission that

[19] See the reviews in *The Athenaeum*, *Literary World*, and *Literary News* for references to spiritualism, and Alfred Taylor Schofield, *Another World, or, The Fourth Dimension* (5th edn., London: Allen and Unwin, 1920), 3–4. Schofield's book was first published in 1888 by Swan Sonnenschein.

beings of a higher order had appeared in and disappeared from Spaceland (p. 105) suggests analogies to Christ's appearances after the resurrection, for instance, and in *The Kernel and the Husk* (p. 259) and *The Spirit on the Waters* (p. 31) Abbott acknowledges that beings from the fourth dimension could produce phenomena that would lend themselves to such explanations. But for Abbott, religion was grounded upon the exercise of Christian virtues, not upon proof for miracles. Even if we could actually conceive of a fourth dimension, 'we should not be a whit the better morally or spiritually' (*Kernel and the Husk*, 259); only the practice of faith, hope, and love can make us better people. Abbott considered spiritualists to be as wrongheaded as Christian fundamentalists for insisting on too literal a proof of the supernatural. Like the Square, who initially assumes that the Sphere must be a deity because of what appear to be his supernatural powers, both groups mistakenly assumed that phenomena that they could not (yet) explain must necessarily have supernatural causes. Similarly, both sides in the empiricist/idealist debate were blinded by their 'respective dimensional prejudices'—by their insistence either that 'This can never be' or that 'It must needs be precisely thus, and we know all about it' (p. 10). In later works Abbott makes explicit the relevance to his own age of the Square's final plea for intellectual modesty about what lies beyond experience. The way to understand the spiritual essence of faith was 'to liberate our thoughts from the yoke of materialism, and to take a more ample view of the Universe', to allow for the possibility that a 'Thoughtland' of the spirit exists which is 'as much more real than Factland as the land of three dimensions seems to us more real than the land of two'.[20] Ultimately he valued non-Euclidean geometries not for the violations of our three-dimensional reality that they allowed, but rather for the higher transcendental realities that they could prepare us to imagine.

[20] Abbott, *Apologia* (London: Black, 1907), 83.

From Flatland to Hyperspace

In addition to the quasi-religious theorizing of scientists like Tait, Stewart, and Zöllner, the popular vogue of the fourth dimension had wide cultural impact in the 1890s and the early decades of the twentieth century. It is exploited as a plot device by authors such as H. G. Wells in 'The Remarkable Case of Davidson's Eyes' (1895), 'The Wonderful Visit' (1895), and 'The Plattner Story' (1896). Higher dimensions also figure in Oscar Wilde's 'The Canterville Ghost' (1891) and Joseph Conrad and Ford Maddox Brown's *The Inheritors* (1901). The protagonist of George MacDonald's *Lilith* (1895) learns he has been preceded in his exploration of higher dimensions by his ancestor 'Sir Upward', a name reminiscent of the Square's attempts to imagine a direction 'Upward yet not Northward' (p. 108). The fourth dimension was employed by Theosophists like P. D. Ouspensky and Charles Webster Leadbeater to help conceptualize the astral projection of the self into higher spiritual worlds. In European visual art, Cubists similarly in search of ways of freeing themselves from the constraints of conventional points of view found support in the existence of a fourth dimension, as artists like Picasso and Braque attempted to transcend Renaissance perspectival conventions by portraying multiple dimensions simultaneously. The most sustained late Victorian meditations on the possibilities of the fourth dimension in this period came from Charles Howard Hinton. Hinton followed up his early essay 'What is the Fourth Dimension?' with a number of other works that extrapolated its implications for other areas. Although in 'A Plane World' (first published in 1884) and 'An Episode of Flatland' (1907) Hinton was more concerned than Abbott with exploring the physical conditions of two-dimensional life,[21] he too saw higher dimensional thinking as a means of promoting spiritual growth. In works like *A New Era of Thought* (1888) and *The Fourth Dimension* (1904) he advocated

[21] In an edition of *Scientific Romances* first published in 1888, Hinton commented that 'evidently the physical conditions of life on the plane have not been . . . [the] main object' of *Flatland*, whereas in his own essay, 'A Plane World', 'we wish, in the first place, to know the physical facts' of two-dimensional life: 'Introduction', *Scientific Romances* (London: Allen and Unwin, 1924), 129.

exercises in conceiving higher dimensions (facilitated by using sets of cubes, sold by his publisher, with multi-coloured sides to represent the different sections of a tesseract or hypercube) as a means of eliminating 'self elements' in our thought and opening our minds to a higher immaterial reality in which all humanity was one. The mental exercise of trying to visualize higher dimensional objects, which one chagrined practitioner of Hinton's method described as 'mind-destroying',[22] is made much easier today by the efforts of mathematicians like Thomas Banchoff, who have developed computer graphics programs that can project images of how higher dimensional objects would appear as they rotate through lower dimensional space.

Abbott's particular meditations on the functioning of two-dimensional and higher geometries have also continued to inspire interest and imitation among later twentieth-century mathematicians. *Sphereland*, Dionys Burger's 1965 sequel to *Flatland*, chronicles the adventures of one of the Square's descendants who discovers that his apparently flat world is actually mapped onto the surface of a sphere. Ian Stewart's *Flatterland* (2001) takes this trope several steps further by sending the Square's granddaughter on a journey through the mysteries of space-time and quantum theory. A. K. Dewdney's *The Planiverse* (1984) further pursues Hinton's investigation of the physics of a two-dimensional world. More fancifully, Rudy Rucker's 1983 short story 'Message Found in a Copy of *Flatland*' imagines what happens when an American academic discovers that the real Flatland still exists on the site of the old City of London School. Rucker's latest homage to Abbott, *Spaceland* (2002), offers a post-modern updating of *Flatland* in the experiences of one 'Joe Cube', who is whisked off into higher dimensional adventures at the beginning of the third millennium.

In the realm of physical science, there had been increasing interest from the mid-nineteenth century in using higher spatial dimensions as a means of solving the most important questions of

[22] See the letter that Martin Gardner prints as an addendum to his chapter on 'Hypercubes' in *Mathematical Carnival* (New York: Knopf, 1975), 52.

physics: the nature of matter and the interrelationship between electricity, magnetism, light, and gravity. As early as the 1850s, German mathematician Bernhard Riemann had challenged the adequacy of Euclidean axioms by positing the possibility of a fourth and higher dimensions, and had argued that the warping of our three-dimensional world into a higher dimensional space could explain forces like gravity, electricity, and magnetism. British mathematician W. K. Clifford, who made Riemann's work available in English translation in 1873, also speculated about the implications of curved space for explaining the substance and motion of matter. In *The Grammar of Science* (1892), Karl Pearson hypothesized that matter might result from ether (the unseen medium that nineteenth-century physicists posited to explain 'action at a distance' like gravitation) squirting into our world from the fourth dimension. Hinton too speculated on the ways the fourth dimension could explain physical phenomena like light (a manifestation of vibration in an unseen fourth dimension) and static electricity (the twisting of matter in the fourth dimension). Just as the Square is forced to admit (p. 7) that the plane figures in Flatland in fact must have some additional three-dimensional 'height' even if it cannot be measured (at least not by Flatland means), Hinton argued that our failure to perceive the fourth dimension could result from the fact that it too is immeasurably small, probably inhering in the smallest particles of matter.

Ultimately, however, the fourth dimension that revolutionized attempts to formulate a unified field theory in the early twentieth century turned out to be not spatial but temporal. H. G. Wells had anticipated some aspects of this development in science fiction like *The Time Machine* (1895), which posited time as the fourth dimension. Albert Einstein's special theory of relativity hypothesized in 1905 that space and time were relative to one another, a relationship given more concrete conceptualization as a four-dimensional space-time continuum in 1907 by mathematician Hermann Minkowski. In his 1916 general theory of relativity, Einstein in effect expanded upon Riemann's earlier theorizing that what we experience as force or gravity could result from the bending of geometrical space by arguing that the presence of matter could cause the bending of space-time, a hypothesis confirmed empirically in 1919, when it was demonstrated during an eclipse that the sun bent the light from nearby stars.

It was not that higher spatial dimensions were completely neglected in scientific thinking during the early twentieth century. The theorizing of Theodor Kaluza in 1919 restored the idea of a fourth spatial dimension (this one in addition to time) by demonstrating how Einstein's field equations for gravity could be reconciled with James Clerk Maxwell's equations for electromagnetism by calculating them in five dimensions instead of Einstein's four. This fifth dimension, like Hinton's fourth, was further theorized by mathematician Oskar Klein in 1926 to be too small for detection in our three-dimensional world. Although the development of quantum mechanics in the 1920s decisively shifted attention away from higher dimensional space for much of the twentieth century, ultimately its failure to develop a unified field theory that satisfactorily incorporated gravity with other physical forces (a goal that eluded Einstein as well) led to a revival of interest in higher spatial dimensions. In the 1980s physicists began to return to theories like those of Kaluza and Klein in an effort to use the perspectives of higher dimensional geometries to explain the functioning of matter and of the primary physical forces of our universe. Such geometries have taken on new importance in contemporary theories of hyperspace, a realm of multiple higher dimensions intersecting with our own. The most promising current model for a 'Theory of Everything' that would unify relativity, gravity, and quantum mechanics is superstring theory, which hypothesizes that what we perceive as the elementary particles of matter are actually the vibration of multi-dimensional strings floating in space-time. Because of its physical manifestations, particularly the symmetries it produces, this space-time is often hypothesized to have ten dimensions, all but four of them infinitesimally small, just as predicted earlier by theorists like Hinton and Klein. Physicists are no more able to perceive these higher dimensions empirically than were nineteenth-century geometers like Flatland's Square, but researchers are engaged in experiments designed to demonstrate the effects of their impact on our lower dimensional world. Such efforts hold out the promise that the cloud-like vision of ten dimensions on *Flatland*'s original cover may prove prophetic, and that we will some day transcend the dimensional limits of our own world and grasp the ultimate truths that structure our universe. Now more than ever can we appreciate Abbott's inspired lessons in how to open our imaginations to what lies beyond the limits of our own experience.

NOTE ON THE TEXT

Flatland was first published in November 1884 in London by Seeley and Company. The 'New and Revised Edition', upon which the Oxford World's Classics is based, followed in December 1884. The December edition added the 'Preface to the Second and Revised Edition, 1884, by the Editor'. In the Preface Abbott responded to two types of criticism by the work's first readers: those concerning the Square's geometrical reasoning, and those concerning his attitudes toward women and the lower classes. As the footnote on p. 9 indicates, in the second edition Abbott also added two further passages to the Square's dialogue with the Sphere in Chapters 16 and 19 to address the first point. He also corrected some errors in the first edition and made some minor changes in wording and punctuation. The 'Third Edition, Revised' was published in Oxford by Basil Blackwell in 1926. It corrected a few additional errors in the second edition, named Abbott on the title page, and added an introduction by William Garnett, a former student of Abbott's. The first American edition was published by Roberts Brothers in Boston early in 1885. With a few exceptions, it followed the first London edition. A comparison of some of the more significant differences among these editions is listed below.

The first and second editions of *Flatland* had an illustrated cover, which is reproduced at the beginning of this edition. The Blackwell third edition also includes this cover illustration. The title page for the Roberts Brothers 1885 edition retained the drawing of clouds from the cover of the first and second editions but eliminated the drawing of the Square's home. It also eliminated the two quotations from *Hamlet* on the original cover ('O day and night, but this is wondrous strange', 'And therefore as a stranger give it welcome') and replaced them with a single quotation from *Titus Andronicus* ('Fie, fie, how franticly I square my talk!'), which had appeared on the title page of the first and second editions. Other twentieth-century editions have for the most part used the Roberts Brothers version of the cover illustration and added 'O day and night, but this is wondrous strange' at the top of the drawing.

	1st edn.	*2nd edn.*	*3rd edn.*
37	*missing the label (1) over the first drawing and the A at the triangle's vertex*	*adds the A*	*adds the (1)*
46	threw a transient charm over	threw a transient splendour over	
53	Dodecahedron	Dodecagon	
63	Dodecahedron	Dodecagon	
80	you in Flatland	you in Spaceland (*2nd edn. and Roberts Brothers edn.*)	
83	second Millennium	second Millennium	third Millennium
84	As soon as the sound of my Wife's retreating footsteps had died away	As soon as the sound of the Peace-cry of my departing Wife had died away	
86–7		*Stranger*. (*To himself*). What must I do? . . . *Stranger*. (*To himself*). I can do neither. *added*	
103–4		*I*. Not inconceivable, my Lord, to me . . . *Sphere*. Analogy! Nonsense: what analogy? *added*	

SELECT BIBLIOGRAPHY

Editions

A Square [Edwin Abbott Abbott], *Flatland: A Romance of Many Dimensions* (London: Seeley, 1884).

—— *Flatland: A Romance of Many Dimensions*, new and rev. edn. (London: Seeley, 1884).

—— *Flatland: A Romance of Many Dimensions*, 3rd edn. rev., with an introduction by William Garnett (Oxford: Basil Blackwell, 1926).

Of the many modern editions of *Flatland*, Ian Stewart's provides the most extensive background information, particularly on mathematical concepts:

Stewart, Ian, *The Annotated Flatland: A Romance of Many Dimensions* (Cambridge, Mass.: Perseus Publishing, 2002).

Biography

To date, no biography of Abbott exists. The most useful sources of background information on his life and work are:

'Abbott, Edwin Abbott', in *Oxford Dictionary of National Biography* (Oxford: Oxford University Press, 2004).

Coonen, Martin, 'Edwin Abbott Abbott: Primary and Secondary Checklists with Partial Annotations', *Bulletin of Bibliography*, 54/4 (Dec. 1999), 247–55.

Douglas-Smith, Aubrey Edward, *The City of London School* (1937), 2nd edn. (Oxford: Basil Blackwell, 1965).

'Obituary. Dr. E. A. Abbott. Scholar, Critic, and Teacher', *The Times* (London), 13 Oct. 1926: 19.

Contemporary Reviews and Notices of Flatland

'Comment and Criticism', *Science*, 5 (1885), 265–6.

'Current Literature', *The Academy*, 8 Nov. 1884: 302.

'Flatland', *Literary World*, 21 Mar. 1885: 93.

'*Flatland*', *Oxford Magazine*, 2 (5 Nov. 1884), 387.

G[arnett], W[illiam], 'Euclid, Newton, and Einstein', *Nature*, 12 Feb. 1920: 627–30.

'Humor and Satire', *Literary News*, NS 6 (Mar. 1885), 85 (reprinted from *Boston Advertiser*, 1885).
'New Books', *New York Times*, 23 Feb. 1885: 3.
'Notes and News', *Science*, 5 (1885), 184.
'Our Library Table', *The Athenaeum*, 15 Nov. 1884: 622.
[Review], *City of London School Magazine*, 8 (1884), 217–21.
[Review], *The Spectator*, 29 Nov. 1884: 1583–4.
Tucker, Robert, 'Flatland', *Nature*, 27 Nov. 1884: 76–7.

Critical Studies of Flatland

Banchoff, Thomas, 'From *Flatland* to Hypergraphics: Interacting with Higher Dimensions', *Interdisciplinary Science Reviews*, 15 (1990), 364–72.
Gilbert, Eliot, ' "Upward, Not Northward": *Flatland* and the Quest for the New', *English Literature in Transition*, 34 (1991), 391–404.
Jann, Rosemary, 'Abbott's *Flatland*: Scientific Imagination and "Natural Christianity" ', *Victorian Studies*, 28 (1985), 473–90.
—— 'Christianity, Spiritualism, and the Fourth Dimension in Late Victorian England', *Victorian Newsletter*, 70 (Fall 1986), 24–8.
Smith, Jonathan, 'Chapter 6: "Euclid Honourably Shelved": Edwin Abbott's *Flatland* and the Methods of Non-Euclidean Geometry', *Fact and Feeling: Baconian Science and the Nineteenth-Century Literary Imagination* (Madison: University of Wisconsin Press, 1994), 180–210.
—— Berkove, Lawrence I., and Baker, Gerald A., 'A Grammar of Dissent: *Flatland*, Newman, and the Theology of Probability', *Victorian Studies*, 39/2 (1996), 129–50.

Background Studies in Mathematics and Higher Dimensions

Banchoff, Thomas F., *Beyond the Third Dimension: Geometry, Computer Graphics, and Higher Dimensions* (New York: Scientific American Library, 1990).
Henderson, Linda Dalrymple, *The Fourth Dimension and Non-Euclidean Geometry in Modern Art* (Princeton: Princeton University Press, 1983).
Kaku, Michio, *Hyperspace: A Scientific Odyssey through Parallel Universes, Time Warps, and the Tenth Dimension* (New York: Oxford University Press, 1994).
Richards, Joan L., *Mathematical Visions: The Pursuit of Geometry in Victorian England* (Boston: Academic Press, 1988).

Background Studies in Victorian Culture and Society

Chadwick, Owen, *The Victorian Church*, 2 vols. (New York: Oxford University Press, 1966–70).

Gilmour, Robin, *The Victorian Period: The Intellectual and Cultural Context of English Literature 1830–1890* (London: Longman, 1993).

Perkin, Harold, *The Origins of Modern English Society 1780–1880* (London: Routledge and Kegan Paul, 1969).

Further Reading in Oxford World's Classics

Carroll, Lewis, *Alice's Adventures in Wonderland/Through the Looking-Glass*, ed. Roger Lancelyn Green.

Darwin, Charles, *The Origin of Species*, ed. Gillian Beer.

Otis, Laura (ed.), *Literature and Science in the Nineteenth Century*.

Smiles, Samuel, *Self-Help*, ed. Peter W. Sinnema.

Swift, Jonathan, *Gulliver's Travels*, ed. Claude Rawson and Ian Higgins.

A CHRONOLOGY OF EDWIN A. ABBOTT

1838	Abbott born 20 December to Edwin Abbott, headmaster of the Philological School, Marylebone, and his wife Jane Abbott Abbott.
1848	Collapse of Chartist movement.
1850–7	Abbott attends City of London School.
1857	Abbott begins university career as a scholarship student at St John's College, Cambridge, where he will win honours as Senior Classic and Senior Chancellor's Medallist.
1859	Charles Darwin, *The Origin of Species*; J. S. Mill, *On Liberty*; Samuel Smiles, *Self-Help*.
1862	Abbott elected Fellow of St John's and ordained deacon in Church of England; appointed assistant headmaster at King Edward's School, Birmingham.
1863	Abbott resigns fellowship to marry Mary Elizabeth Rangeley; ordained priest.
1864	Abbott appointed second master at Clifton College.
1865	Abbott becomes headmaster of City of London School.
	Lewis Carroll, *Alice's Adventures in Wonderland*.
1867	Second Reform Act.
1869	Abbott, *A Shakespearean Grammar*.
	Founding of Girton College for women; Francis Galton, *Hereditary Genius*; J. S. Mill, *The Subjection of Women*.
1870	Abbott, *Bible Lessons*.
	Education Act authorizes publicly supported elementary schools; Hermann von Helmholtz, 'The Axioms of Geometry'.
1871	Abbott, *English Lessons for English People* with J. R. Seeley.
	Lewis Carroll, *Through the Looking Glass*.
1872	Abbott, *How to Write Clearly*; *The Good Voices: A Child's Guide to the Bible*; attends as platform guest at founding of Girls' Public Day School Company.
	Charles Darwin, *The Descent of Man*; Samuel Butler, *Erewhon*.

1873 Abbott, *Latin Prose through English Idiom*; *Parables for Children*.

1874 Abbott, *How to Tell the Parts of Speech*; *Handbook of English Grammar*.

1875 Abbott, *Cambridge Sermons, Preached before the University*; 'Gospels' in *Encyclopedia Britannica*; invited to be select preacher at Oxford.

P. G. Tait and Balfour Stewart, *The Unseen Universe*.

1876 Abbott, edition of *Bacon's Essays*; elected Hulsean Lecturer at Cambridge.

1877 Abbott, *Bacon and Essex: A Sketch of Bacon's Earlier Life*; *Through Nature to Christ, or, The Ascent of Worship Through Illusion to the Truth*.

1878 Abbott, *Philochristus: Memoirs of a Disciple of the Lord* (published anonymously).

P. G. Tait and Balfour Stewart, *Paradoxical Philosophy*.

1879 Abbott, *Oxford Sermons, Preached before the University*.

1880 Charles Howard Hinton, 'What is the Fourth Dimension?'; Friedrich Zöllner, *Transcendental Physics*.

1882 Abbott, *Onesimus: Memoirs of a Disciple of Paul* (published anonymously).

1883 Abbott, *Hints on Home Teaching*.

1884 Abbott, *The Common Tradition of the Synoptic Gospels in the Text of the Revised Version* with W. G. Rushbrooke; *Flatland: A Romance of Many Dimensions* (published anonymously).

Third Reform Act.

1885 Abbott, *Francis Bacon: An Account of his Life and Works*.

1886 Abbott, *Via Latina: A First Latin Book*; *The Kernel and the Husk: Letters on Spiritual Christianity* (published anonymously).

Charles Howard Hinton, *Scientific Romances, First Series*.

1888 Charles Howard Hinton, *A New Era of Thought*.

1889 Abbott, *The Latin Gate: A First Latin Translation Book*; resigns as headmaster of City of London School.

1891 Abbott, *Philomythus, an Antidote against Credulity: A Discussion of Cardinal Newman's Essay on Ecclesiastical*

	Miracles; *Newmanianism: A Preface to the Second Edition of Philomythus*.
1892	Abbott, *The Anglican Career of Cardinal Newman*.
1893	Abbott, *Dux Latinus: A First Latin Construing Book*.
1895	H. G. Wells, *The Time Machine*; George MacDonald, *Lilith*.
1897	Abbott, *The Spirit on the Waters: The Evolution of the Divine from the Human*.
1898	Abbott, *St Thomas of Canterbury: His Death and Miracles*.
1900	Abbott, *Clue: A Guide Through Greek to Hebrew Scripture*.
1901	Abbott, *The Corrections of Mark: Adopted by Matthew and Luke*.
	Death of Queen Victoria; succeeded by Edward VII.
1902	Charles Howard Hinton, *Scientific Romances, Second Series*.
1903	Abbott, *Contrast; or, A Prophet and a Forger*; *From Letter to Spirit: An Attempt to Reach Through Varying Voices the Abiding Word*.
1904	Abbott, *Paradosis, or, 'In the Night in Which He Was (?) Betrayed'*.
	Charles Howard Hinton, *The Fourth Dimension*.
1905	Abbott, *Johannine Vocabulary: A Comparison of the Words of the Fourth Gospel with Those of the Three*.
	Albert Einstein, Special Theory of Relativity.
1906	Abbott, *Silanus the Christian*; *Johannine Grammar*.
1907	Abbott, *Apologia: An Explanation and Defence*; *Notes on New Testament Criticism*; *Indices to Diatessarica*.
	Charles Howard Hinton, *An Episode of Flatland*; Hermann Minkowski, *Space and Time*.
1909	Abbott, *The Message of the Son of Man*.
1910	Abbott, *The Son of Man; or, Contributions to the Study of the Thoughts of Jesus*.
	Death of King Edward; succeeded by George V.
1912	Abbott, *Light on the Gospel from an Ancient Poet*; elected Honorary Fellow of St John's College, Cambridge.
1913	Abbott, *The Fourfold Gospel, Section I*; *Miscellanea Evangelica*.

1914	Abbott, *The Fourfold Gospel, Section II*.
	Start of First World War.
1915	Abbott, *The Fourfold Gospel, Section III*.
1916	Abbott, *The Fourfold Gospel, Section IV*.
	Albert Einstein, General Theory of Relativity.
1917	Abbott, *The Fourfold Gospel, Section V*.
1918	End of First World War.
1926	Abbott dies 12 October; buried in Hampstead Cemetery.

"O day and night, but this is wondrous strange"

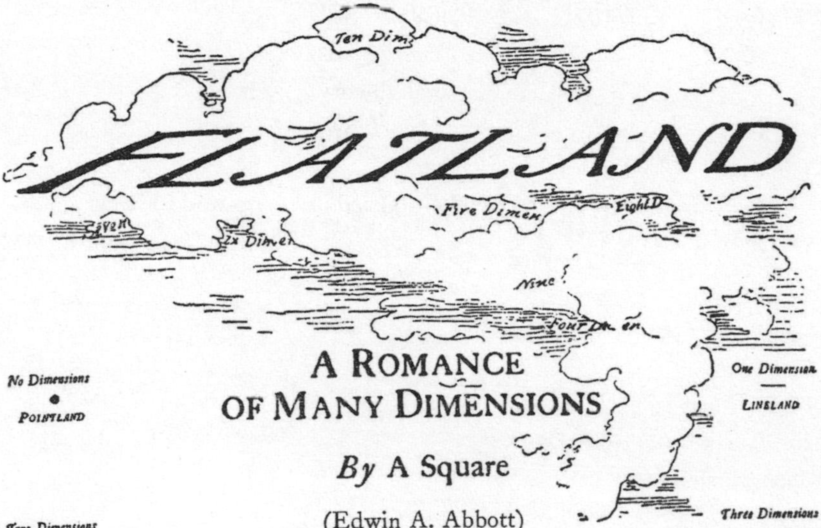

FLATLAND

A ROMANCE OF MANY DIMENSIONS

By A Square

(Edwin A. Abbott)

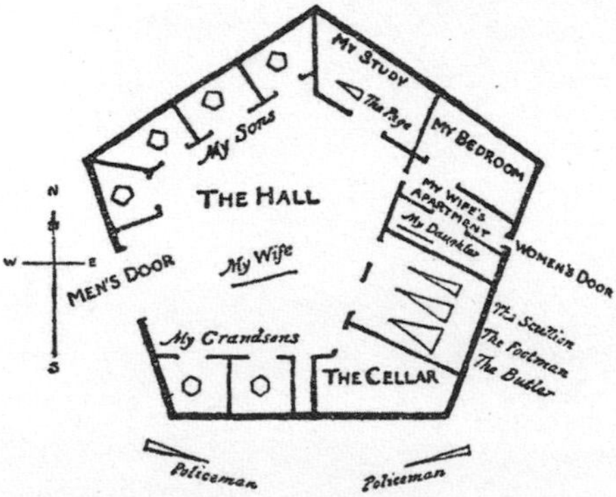

" And therefore as a stranger give it welcome."

BASIL BLACKWELL · OXFORD

Price Seven Shillings and Sixpence net.

FLATLAND

A Romance of Many Dimensions

With Illustrations
by the Author, A SQUARE

'Fie, fie, how franticly I square my talk!'

To

The Inhabitants of SPACE IN GENERAL

And H. C. IN PARTICULAR*

This Work is Dedicated

By a Humble Native of Flatland

In the Hope that

Even as he was Initiated into the Mysteries

Of THREE Dimensions

Having been previously conversant

With ONLY TWO

So the Citizens of that Celestial Region

May aspire yet higher and higher

To the Secrets of FOUR FIVE OR EVEN SIX Dimensions

Thereby contributing

To the Enlargement of THE IMAGINATION

And the possible Development

Of that most rare and excellent Gift of MODESTY

Among the Superior Races

Of SOLID HUMANITY

PREFACE TO THE SECOND AND REVISED EDITION

BY THE EDITOR*

IF my poor Flatland friend retained the vigour of mind which he enjoyed when he began to compose these Memoirs, I should not now need to represent him in this Preface, in which he desires, firstly, to return his thanks to his readers and critics in Spaceland, whose appreciation has, with unexpected celerity, required a second edition of his work; secondly, to apologize for certain errors and misprints (for which, however, he is not entirely responsible); and, thirdly, to explain one or two misconceptions. But he is not the Square he once was. Years of imprisonment, and the still heavier burden of general incredulity and mockery, have combined with the natural decay of old age to erase from his mind many of the thoughts and notions, and much also of the terminology, which he acquired during his short stay in Spaceland. He has, therefore, requested me to reply in his behalf to two special objections, one of an intellectual, the other of a moral nature.

The first objection is, that a Flatlander, seeing a Line, sees something that must be *thick* to the eye as well as *long* to the eye (otherwise it would not be visible, if it had not some thickness); and consequently he ought (it is argued) to acknowledge that his countrymen are not only long and broad, but also (though doubtless in a very slight degree) *thick* or *high*. This objection is plausible, and, to Spacelanders, almost irresistible, so that, I confess, when I first heard it, I knew not what to reply. But my poor old friend's answer appears to me completely to meet it.

'I admit,' said he—when I mentioned to him this objection—'I admit the truth of your critic's facts, but I deny his conclusions. It is true that we have really in Flatland a Third unrecognized Dimension called "height," just as it is also true that you have really in Spaceland a Fourth unrecognized Dimension, called by no name at present, but which I will call "extra-height." But we can no more take cognizance of our "height" than you can of your "extra-height." Even I—who have been in Spaceland, and have had the privilege of understanding for twenty-four hours the meaning of "height"—even I cannot now comprehend it, nor realise it by the sense of sight or by any process of reason: I can but apprehend it by faith.

'The reason is obvious. Dimension implies direction, implies measurement, implies the more and the less. Now, all our lines are *equally* and *infinitesimally* thick (or high, whichever you like); consequently, there is nothing in them to lead our minds to the conception of that Dimension. No "delicate micrometer"—as has been suggested by one too hasty Spaceland critic—would in the least avail us; for we should not know *what to measure, nor in what direction*. When we see a Line, we see something that is long and *bright*; *brightness*, as well as length, is necessary to the existence of a Line; if the brightness vanishes, the Line is extinguished. Hence, all my Flatland friends—when I talk to them about the unrecognized Dimension which is somehow visible in a Line—say, "Ah, you mean *brightness*": and when I reply, "No, I mean a real Dimension," they at once retort, "Then measure it; or tell us in what direction it extends": and this silences me, for I can do neither. Only yesterday, when the Chief Circle (in other words our High Priest) came to inspect the State Prison and paid me his seventh annual visit, and when for the seventh time he put me the question, "Was I any better?" I tried to prove to him that he was "high," as well as long and broad, although he did not know it. But what was his reply? "You say I am 'high'; measure my 'high-ness' and I will believe you." What could I do? How could I meet his challenge? I was crushed; and he left the room triumphant.

'Does this still seem strange to you? Then put yourself in a similar position. Suppose a person of the Fourth Dimension, condescending to visit you, were to say, "Whenever you open your eyes, you *see* a Plane (which is of Two Dimensions) and you *infer* a Solid (which is of Three); but in reality you also see (though you do not recognize) a Fourth Dimension, which is not colour nor brightness nor anything of the kind, but a true Dimension, although I cannot point out to you its direction, nor can you possibly measure it." What would you say to such a visitor? Would not you have him locked up? Well, that is my fate: and it is as natural for us Flatlanders to lock up a Square for preaching the Third Dimension, as it is for you Spacelanders to lock up a Cube for preaching the Fourth. Alas, how strong a family likeness runs through blind and persecuting humanity in all Dimensions! Points, Lines, Squares, Cubes, Extra-Cubes*—we are all liable to the same errors, all alike the Slaves of our respective Dimensional prejudices, as one of your own Spaceland poets has said—

"One touch of Nature makes all worlds akin."[1]*

On this point the defence of the Square seems to me to be impregnable. I wish I could say that his answer to the second (or moral) objection was equally clear and cogent. It has been objected that he is a woman-hater; and as this objection has been vehemently urged by those whom Nature's decree has constituted the somewhat larger half of the Spaceland race, I should like to remove it, so far as I can honestly do so. But the Square is so unaccustomed to the use of the moral terminology of Spaceland that I should be doing him an injustice if I were literally to transcribe his defence against this charge. Acting, therefore, as his interpreter and summarizer, I gather that in the course of an imprisonment of seven years he has himself modified his own personal views, both as regards Women and as regards the Isosceles or Lower Classes. Personally, he now inclines to the opinion of the Sphere (see pages 97–8) that the Straight Lines are in many important respects superior to the Circles. But, writing as a Historian, he has identified himself (perhaps too closely) with the views generally adopted by Flatland, and (as he has been informed) even by Spaceland, Historians; in whose pages (until very recent times) the destinies of Women and of the masses of mankind have seldom been deemed worthy of mention and never of careful consideration.

[1] The Author desires me to add, that the misconception of some of his critics on this matter has induced him to insert (on pages 86–7 and 103–4) in his dialogue with the Sphere, certain remarks which have a bearing on the point in question, and which he had previously omitted as being tedious and unnecessary.

In a still more obscure passage he now desires to disavow the Circular or aristocratic tendencies with which some critics have naturally credited him. While doing justice to the intellectual power with which a few Circles have for many generations maintained their supremacy over immense multitudes of their countrymen, he believes that the facts of Flatland, speaking for themselves without comment on his part, declare that Revolutions cannot always be suppressed by slaughter, and that Nature, in sentencing the Circles to infecundity, has condemned them to ultimate failure—'and herein,' he says, 'I see a fulfilment of the great Law of all worlds, that while the wisdom of Man thinks it is working one thing, the wisdom of Nature constrains it to work another, and quite a different and far better, thing.' For the rest, he begs his readers not to suppose that every minute detail in the daily life of Flatland must needs correspond to some other detail in Spaceland; and yet he hopes that, taken as a whole, his work may prove suggestive, as well as amusing, to those Spacelanders of moderate and modest minds who—speaking of that which is of the highest importance, but lies beyond experience—decline to say on the one hand, 'This can never be,' and on the other hand, 'It must needs be precisely thus, and we know all about it.'

CONTENTS

PART I
THIS WORLD

PART II
OTHER WORLDS

PART I

THIS WORLD

'Be patient, for the world is broad and wide.' *

I

Of the Nature of Flatland

I CALL our world Flatland, not because we call it so, but to make its nature clearer to you, my happy readers, who are privileged to live in Space.

Imagine a vast sheet of paper on which straight Lines, Triangles, Squares, Pentagons, Hexagons, and other figures, instead of remaining fixed in their places, move freely about, on or in the surface, but without the power of rising above or sinking below it, very much like shadows—only hard and with luminous edges—and you will then have a pretty correct notion of my country and countrymen. Alas, a few years ago, I should have said 'my universe': but now my mind has been opened to higher views of things.

In such a country, you will perceive at once that it is impossible that there should be anything of what you call a 'solid' kind; but I dare say you will suppose that we could at least distinguish by sight the Triangles, Squares, and other figures, moving about as I have described them. On the contrary, we could see nothing of the kind, not at least so as to distinguish one figure from another. Nothing was visible, nor could be visible, to us, except straight Lines; and the necessity of this I will speedily demonstrate.

Place a penny on the middle of one of your tables in Space; and leaning over it, look down upon it. It will appear a circle.

But now, drawing back to the edge of the table, gradually lower your eye (thus bringing yourself more and more into the condition of the inhabitants of Flatland), and you will find the penny becoming more and more oval to your view; and at last when you have placed your eye exactly on the edge of the table (so that you are, as it were, actually a Flatlander) the penny will then have ceased to appear oval at all, and will have become, so far as you can see, a straight line.

The same thing would happen if you were to treat in the same way a Triangle, or Square, or any other figure cut out of pasteboard. As soon as you look at it with your eye on the edge of the table, you will find that it ceases to appear to you a figure, and that it becomes in appearance a straight line. Take for example an equilateral Triangle—who represents with us a Tradesman of the respectable class. Fig. 1 represents the Tradesman as you would see him while you were bending over him from above; figs. 2 and 3 represent the Tradesman, as you would see him if your eye were close to the level, or all but on the level of the table; and if your eye were quite on the level of the table (and that is how we see him in Flatland) you would see nothing but a straight line.

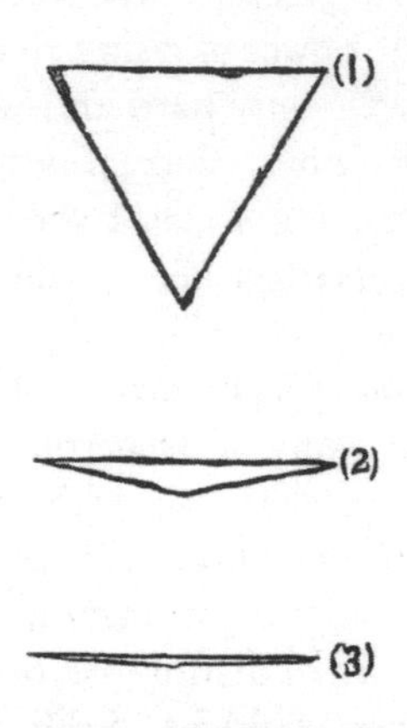

When I was in Spaceland I heard that your sailors have very similar experiences while they traverse your seas and discern some distant island or coast lying on the horizon. The far-off land may have bays, forelands, angles in and out to any number and extent; yet at a distance you see none of these (unless indeed your sun shines bright upon them, revealing the projections and retirements by means of light and shade), nothing but a grey unbroken line upon the water.

Well, that is just what we see when one of our triangular or other acquaintances comes towards us in Flatland. As there is neither sun with us, nor any light of such a kind as to make shadows, we have none of the helps to the sight that you have in Spaceland. If our friend comes close to us we see his line becomes larger; if he leaves us it becomes smaller: but still he looks like a straight line; be he a Triangle, Square, Pentagon, Hexagon, Circle, what you will—a straight Line he looks and nothing else.

You may perhaps ask how under these disadvantageous circumstances we are able to distinguish our friends from one another: but the answer to this very natural question will be more fitly and easily given when I come to describe the inhabitants of Flatland. For the present let me defer this subject, and say a word or two about the climate and houses in our country.

2

Of the Climate and Houses in Flatland

As with you, so also with us, there are four points of the compass North, South, East, and West.

There being no sun nor other heavenly bodies, it is impossible for us to determine the North in the usual way; but we have a method of our own. By a Law of Nature with us, there is a constant attraction to the South; and, although in temperate climates this is very slight—so that even a Woman in reasonable health can journey several furlongs northward without much difficulty—yet the hampering effect of the southward attraction is quite sufficient to serve as a compass in most parts of our earth. Moreover the rain (which falls at stated intervals) coming always from the North, is an additional assistance; and in the towns we have the guidance of the houses, which of course have their side-walls running for the most part North and South, so that the roofs may keep off the rain from the North. In the country, where there are no houses, the trunks of the trees serve as some sort of guide. Altogether, we have not so much difficulty as might be expected in determining our bearings.

Yet in our more temperate regions, in which the southward attraction is hardly felt, walking sometimes in a perfectly desolate plain where there have been no houses nor trees to guide me, I have been occasionally compelled to remain stationary for hours together, waiting till the rain came before continuing my journey. On the weak and aged, and especially on delicate Females, the force of attraction tells much more heavily than on the robust of the Male Sex, so that it is a point of breeding, if you meet a Lady in the street always to give her the North side of the way*—by no means an easy thing to do always at short notice when you are in rude health and in a climate where it is difficult to tell your North from your South.

Windows there are none in our houses: for the light comes to us alike in our homes and out of them, by day and by night, equally at all times and in all places, whence we know not. It was in old days, with our learned men, an interesting and oft-investigated question, 'What is the origin of light?' and the solution of it has been repeatedly attempted, with no other result than to crowd our lunatic asylums with the would-be solvers. Hence, after fruitless attempts to suppress such investigations indirectly by making them liable to a heavy tax, the Legislature, in comparatively recent times, absolutely prohibited them. I—alas I alone in Flatland—know now only too well the true solution of this mysterious problem; but my knowledge cannot be made intelligible to a single one of my countrymen; and I am mocked at—I, the sole possessor of the truths of Space and of the theory of the introduction of Light from the world of Three Dimensions—as if I were the maddest of the mad! But a truce of these painful digressions: let me return to our houses.

The most common form for the construction of a house is five-sided or pentagonal, as in the annexed figure. The two Northern sides *RO*, *OF*, constitute the roof, and for the most part have no doors; on the East is a small door for the Women; on the West a much larger one for the Men; the South side or floor is usually doorless.

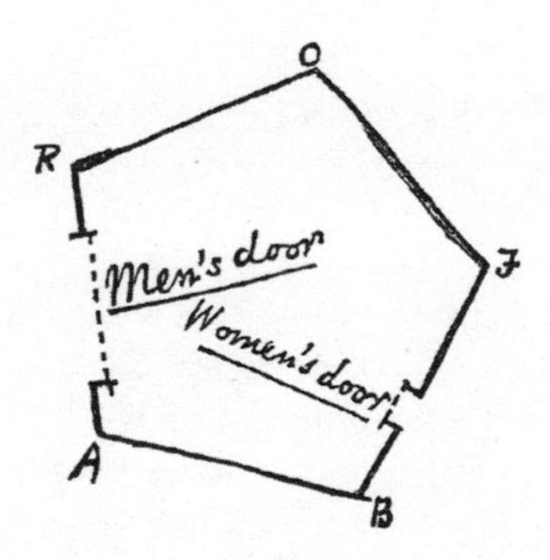

Square and triangular houses are not allowed, and for this reason. The angles of a Square (and still more those of an equilateral Triangle) being much more pointed than those of a Pentagon, and the lines of inanimate objects (such as houses) being dimmer than the lines of Men and Women, it follows that there is no little danger lest the points of a square or triangular house residence might do serious injury to an inconsiderate or perhaps absent-minded traveller suddenly running against them: and therefore, as early as the eleventh century of our era, triangular houses were universally forbidden by Law, the only exceptions being fortifications, powder-magazines, barracks, and other state buildings, which it is not desirable that the general public should approach without circumspection.

At this period, square houses were still everywhere permitted, though discouraged by a special tax. But, about three centuries afterwards, the Law decided that in all towns containing a population above ten thousand, the angle of a Pentagon was the smallest house-angle that could be allowed consistently with the public safety. The good sense of the community has seconded the efforts of the Legislature; and now, even in the country, the pentagonal construction has superseded every other. It is only now and then in some very remote and backward agricultural district that an antiquarian may still discover a square house.

3

Concerning the Inhabitants of Flatland

THE greatest length or breadth of a full-grown inhabitant of Flatland may be estimated at about eleven of your inches. Twelve inches may be regarded as a maximum.

Our Women are Straight Lines.

Our Soldiers and Lowest Classes of Workmen are Triangles with two equal sides, each about eleven inches long, and a base or third side so short (often not exceeding half an inch) that they form at their vertices a very sharp and formidable angle. Indeed when their bases are of the most degraded type (not more than the eighth part of an inch in size) they can hardly be distinguished from Straight Lines or Women; so extremely pointed are their vertices. With us, as with you, these Triangles are distinguished from others by being called Isosceles; and by this name I shall refer to them in the following pages.

Our Middle Class consists of Equilateral or Equal-sided Triangles.

Our Professional Men and Gentlemen* are Squares (to which class I myself belong) and Five-sided figures or Pentagons.

Next above these come the Nobility, of whom there are several degrees, beginning at Six-sided Figures, or Hexagons, and from thence rising in the number of their sides till they receive the honourable title of Polygonal, or many-sided. Finally when the number of the sides becomes so numerous, and the sides themselves so small, that the figure cannot be distinguished from a circle, he is included in the Circular or Priestly order; and this is the highest class of all.

It is a Law of Nature* with us that a male child shall have one more side than his father, so that each generation shall rise (as a rule) one step in the scale of development and nobility. Thus the son of a Square is a Pentagon; the son of a Pentagon, a Hexagon; and so on.

But this rule applies, not always to the Tradesmen, and still less often to the Soldiers, and to the Workmen; who indeed can hardly be said to deserve the name of human Figures, since they have not all their sides equal. With them therefore the Law of Nature does not hold; and the son of an Isosceles (*i.e.* a Triangle with two sides equal) remains Isosceles still. Nevertheless, all hope is not shut out, even from the Isosceles, that his posterity may ultimately rise above his degraded condition. For, after a long series of military successes, or diligent and skilful labours, it is generally found that the more intelligent among the Artisan and Soldier classes manifest a slight increase of their third side or base, and a shrinkage of the two other sides. Intermarriages (arranged by the Priests) between the sons and daughters of these more intellectual members of the lower classes generally result in an offspring approximating still more to the type of the Equal-sided Triangle.

Rarely—in proportion to the vast number of Isosceles births—is a genuine and certifiable Equal-sided Triangle produced from Isosceles parents.[1] Such a birth requires, as its antecedents, not only a series of carefully arranged intermarriages, but also a long-continued exercise of frugality and self-control on the part of the would-be ancestors of the coming Equilateral, and a patient, systematic, and continuous development of the Isosceles intellect through many generations.

The birth of a True Equilateral Triangle from Isosceles parents is the subject of rejoicing in our country for many furlongs round. After a strict examination conducted by the Sanitary and Social Board, the infant, if certified as Regular, is with solemn ceremonial admitted into the class of Equilaterals. He is then immediately taken from his proud yet sorrowing parents and

[1] 'What need of a certificate?' a Spaceland critic may ask: 'Is not the procreation of a Square Son a certificate from Nature herself, proving the Equal-sidedness of the Father?' I reply that no Lady of any position will marry an uncertified Triangle. Square offspring has sometimes resulted from a slightly Irregular Triangle: but in almost every such case the Irregularity of the first generation is visited on the third; which either fails to attain the Pentagonal rank, or relapses to the Triangular.

adopted by some childless Equilateral, who is bound by oath never to permit the child henceforth to enter his former home or so much as to look upon his relations again, for fear lest the freshly developed organism may, by force of unconscious imitation, fall back again into his hereditary level.

The occasional emergence of an Isosceles from the ranks of his serf-born ancestors, is welcomed not only by the poor serfs themselves, as a gleam of light and hope shed upon the monotonous squalor of their existence, but also by the Aristocracy at large; for all the higher classes are well aware that these rare phenomena, while they do little or nothing to vulgarise their own privileges, serve as a most useful barrier against revolution from below.

Had the acute-angled rabble been all, without exception, absolutely destitute of hope and of ambition, they might have found leaders in some of their many seditious outbreaks, so able as to render their superior numbers and strength too much for the wisdom even of the Circles. But a wise ordinance of Nature has decreed that, in proportion as the working-classes increase in intelligence, knowledge, and all virtue, in that same proportion their acute angle (which makes them physically terrible) shall increase also and approximate to the harmless angle of the Equilateral Triangle. Thus, in the most brutal and formidable of the soldier class creatures almost on a level with women in their lack of intelligence—it is found that, as they wax in the mental ability necessary to employ their tremendous penetrating power to advantage, so do they wane in the power of penetration itself.

How admirable is this Law of Compensation! And how perfect a proof of the natural fitness and, I may almost say, the divine origin of the aristocratic constitution of the States in Flatland! By a judicious use of this Law of Nature, the Polygons and Circles are almost always able to stifle sedition in its very cradle, taking advantage of the irrepressible and boundless hopefulness of the human mind. Art also comes to the aid of Law and Order. It is generally found possible—by a little artificial compression or expansion on the part of the State physicians—to make some of the more intelligent leaders of a rebellion perfectly Regular, and to admit them at once into the privileged classes; a much larger number, who are still below the standard, allured by the prospect of being ultimately ennobled, are induced to enter the State Hospitals, where they are kept in honourable confinement for life; one or two alone of the more obstinate, foolish, and hopelessly irregular are led to execution.

Then the wretched rabble of the Isosceles, planless and leaderless, are either transfixed without resistance by the small body of their brethren whom the Chief Circle keeps in pay for emergencies of this kind; or else, more often, by means of jealousies and suspicions skilfully fomented among them by the Circular party, they are stirred to mutual warfare, and perish by one another's angles. No less than one hundred and twenty rebellions are recorded in our annals, besides minor outbreaks numbered at two hundred and thirty-five; and they have all ended thus.

4
Concerning the Women

IF our highly pointed Triangles of the Soldier class are formidable, it may be readily inferred that far more formidable are our Women. For, if a Soldier is a wedge, a Woman is a needle; being, so to speak, *all* point, at least at the two extremities. Add to this the power of making herself practically invisible at will, and you will perceive that a Female in Flatland, is a creature by no means to be trifled with.

But here, perhaps, some of my younger Readers may ask *how* a woman in Flatland can make herself invisible. This ought, I think, to be apparent without any explanation. However, a few words will make it clear to the most unreflecting.

Place a needle on a table. Then, with your eye on the level of the table, look at it side-ways, and you see the whole length of it; but look at it end-ways, and you see nothing but a point: it has become practically invisible. Just so is it with one of our Women. When her side is turned towards us, we see her as a straight line; when the end containing her eye or mouth—for with us these two organs are identical—is the part that meets our eye, then we see nothing but a highly lustrous point; but when the back is presented to our view, then—being only sub-lustrous, and, indeed, almost as dim as an inanimate object—her hinder extremity serves her as a kind of Invisible Cap.*

The dangers to which we are exposed from our Women must now be manifest to the meanest capacity in Spaceland. If even the angle of a respectable Triangle in the middle class is not without its dangers; if to run against a Working Man involves a gash; if collision with an Officer of the military class necessitates a serious wound; if a mere touch from the vertex of a Private Soldier brings with it danger of death;—what can it be to run against a Woman, except absolute and immediate destruction? And when a

Woman is invisible, or visible only as a dim sub-lustrous point, how difficult must it be, even for the most cautious, always to avoid collision!

Many are the enactments made at different times in the different States of Flatland, in order to minimize this peril; and in the Southern and less temperate climates, where the force of gravitation is greater, and human beings more liable to casual and involuntary motions, the Laws concerning Women are naturally much more stringent. But a general view of the Code may be obtained from the following summary:—

1. Every house shall have one entrance in the Eastern side, for the use of Females only; by which all Females shall enter 'in a becoming and respectful manner'[1] and not by the Men's or Western door.

2. No Female shall walk in any public place without continually keeping up her Peace-cry* under penalty of death.

3. Any Female, duly certified to be suffering from St. Vitus's Dance,* fits, chronic cold accompanied by violent sneezing, or any disease necessitating involuntary motions, shall be instantly destroyed.

In some of the States there is an additional Law forbidding Females, under penalty of death, from walking or standing in any public place without moving their backs constantly from right to left so as to indicate their presence to those behind them; others oblige a Woman, when travelling, to be followed by one of her sons, or servants, or by her husband; others confine Women altogether to their houses except during the religious festivals. But it has been found by the wisest of our Circles or Statesmen that the multiplication of restrictions on Females tends not only to the debilitation and diminution of the race, but also to the increase of domestic murders to such an extent that a State loses more than it gains by a too prohibitive Code.

For whenever the temper of the Women is thus exasperated by

[1] When I was in Spaceland I understood that some of your Priestly Circles have in the same way a separate entrance for Villagers, Farmers, and Teachers of Board Schools* (*Spectator*, Sept. 1884, p. 1255) that they may 'approach in a becoming and respectful manner.'

confinement at home or hampering regulations abroad, they are apt to vent their spleen upon their husbands and children; and in the less temperate climates the whole male population of a village has been sometimes destroyed in one or two hours of simultaneous female outbreak. Hence the Three Laws, mentioned above, suffice for the better regulated States, and may be accepted as a rough exemplification of our Female Code.

After all, our principal safeguard is found, not in Legislature, but in the interests of the Women themselves. For, although they can inflict instantaneous death by a retrograde movement, yet unless they can at once disengage their stinging extremity from the struggling body of their victim, their own frail bodies are liable to be shattered.

The power of Fashion is also on our side. I pointed out that in some less civilised States no female is suffered to stand in any public place without swaying her back from right to left. This practice has been universal among ladies of any pretensions to breeding in all well-governed States, as far back as the memory of Figures can reach. It is considered a disgrace to any State that legislation should have to enforce what ought to be, and is in every respectable female, a natural instinct. The rhythmical and, if I may so say, well-modulated undulation of the back in our ladies of Circular rank* is envied and imitated by the wife of a common Equilateral, who can achieve nothing beyond a mere monotonous swing, like the ticking of a pendulum; and the regular tick of the Equilateral is no less admired and copied by the wife of the progressive and aspiring Isosceles, in the females of whose family no 'back-motion' of any kind has become as yet a necessity of life. Hence, in every family of position and consideration, 'back motion' is as prevalent as time itself; and the husbands and sons in these households enjoy immunity at least from invisible attacks.

Not that it must be for a moment supposed that our Women are destitute of affection. But unfortunately the passion of the moment predominates, in the Frail Sex, over every other consideration. This is, of course, a necessity arising from their unfortunate conformation. For as they have no pretensions to an angle, being inferior in this respect to the very lowest of the Isosceles, they are consequently wholly devoid of brain-power, and have neither reflection, judgment, nor forethought, and hardly any memory. Hence, in their fits of fury, they remember no claims and recognise no distinctions. I have actually known a case where a Woman has exterminated her whole household, and half an hour afterwards, when her rage was over and the fragments swept away, has asked what has become of her husband and her children!

Obviously then a Woman is not to be irritated as long as she is in a position where she can turn round. When you have them in their apartments—which are constructed with a view to denying them that power—you can say and do what you like; for they are then wholly impotent for mischief, and will not remember a few minutes hence the incident for which they may be at this moment threatening you with death, nor the promises which you may have found it necessary to make in order to pacify their fury.

On the whole we get on pretty smoothly in our domestic relations, except in the lower strata of the Military Classes. There the want of tact and discretion on the part of the husbands produces at times indescribable disasters. Relying too much on the offensive weapons of their acute angles instead of the defensive organs of good sense and seasonable simulations,* these reckless creatures too often neglect the prescribed construction of the Women's apartments, or irritate their wives by ill-advised expressions out of doors, which they refuse immediately to retract. Moreover a blunt and stolid regard for literal truth indisposes them to make those lavish promises by which the more judicious Circle can in a moment pacify his consort. The result is massacre; not however without its advantages, as it eliminates the more brutal and troublesome of the Isosceles; and by many of our Circles the destructiveness of the Thinner Sex is regarded as one among many providential arrangements for suppressing redundant population, and nipping Revolution in the bud.

Yet even in our best regulated and most approximately circular families I cannot say that the ideal of family life is so high as with you in Spaceland. There is peace, in so far as the absence of slaughter may be called by that name, but there is necessarily little harmony of tastes or pursuits; and the cautious wisdom of the Circles has ensured safety at the cost of domestic comfort. In every Circular or Polygonal household it has been a habit from time immemorial—and has now become a kind of instinct among the women of our higher classes—that the mothers and daughters should constantly keep their eyes and mouths towards their husband and his male friends; and for a lady in a family of distinction to turn her back upon her husband would be regarded as a kind of portent, involving loss of *status*. But, as I shall soon shew, this custom, though it has the advantage of safety, is not without its disadvantages.

In the house of the Working Man or respectable Tradesman—where the wife is allowed to turn her back upon her husband, while pursuing her household avocations—there are at least intervals of quiet, when the wife is neither seen nor heard, except for the humming sound of the continuous Peace-cry; but in the homes of the upper classes there is too often no peace. There the voluble mouth and bright penetrating eye are ever directed towards the Master of the household; and light itself is not more persistent than the stream of feminine discourse. The tact and skill which suffice to avert a Woman's sting are unequal to the task of stopping a Woman's mouth; and as the wife has absolutely nothing to say, and absolutely no constraint of wit, sense, or conscience to prevent her from saying it, not a few cynics have been found to aver that they prefer the danger of the death-dealing but inaudible sting to the safe sonorousness of a Woman's other end.

To my readers in Spaceland the condition of our Women may seem truly deplorable, and so indeed it is. A Male of the lowest type of the Isosceles may look forward to some improvement of his angle, and to the ultimate elevation of the whole of his degraded caste; but no Woman can entertain such hopes for her sex. 'Once a Woman, always a Woman' is a Decree of Nature; and the very Laws of Evolution* seem suspended in her disfavour. Yet at least we can admire the wise Prearrangement which has ordained that, as they have no hopes, so they shall have no memory to recall, and no forethought to anticipate, the miseries and humiliations which are at once a necessity of their existence and the basis of the constitution of Flatland.

5

Of our Methods of Recognizing one another

YOU, who are blessed with shade as well as light, you who are gifted with two eyes, endowed with a knowledge of perspective, and charmed with the enjoyment of various colours, you, who can actually *see* an angle, and contemplate the complete circumference of a Circle in the happy region of Three Dimensions—how shall I make clear to you the extreme difficulty which we in Flatland experience in recognizing one another's configuration?

Recall what I told you above. All beings in Flatland, animate or inanimate, no matter what their form, present *to our view* the same, or nearly the same, appearance, viz. that of a straight Line. How then can one be distinguished from another, where all appear the same?

The answer is threefold. The first means of recognition is the sense of hearing; which with us is far more highly developed than with you, and which enables us not only to distinguish by the voice our personal friends, but even to discriminate between different classes, at least so far as concerns the three lowest orders, the Equilateral, the Square, and the Pentagon—for of the Isosceles I take no account. But as we ascend in the social scale, the process of discriminating and being discriminated by hearing increases in difficulty, partly because voices are assimilated, partly because the faculty of voice-discrimination is a plebeian virtue not much developed among the Aristocracy. And wherever there is any danger of imposture we cannot trust to this method. Amongst our lowest orders, the vocal organs are developed to a degree more than correspondent with those of hearing, so that an Isosceles can easily feign the voice of a Polygon, and, with some training, that of a Circle himself. A second method is therefore more commonly resorted to.

Feeling is, among our Women and lower classes—about

our upper classes I shall speak presently—the principal test of recognition, at all events between strangers, and when the question is, not as to the individual, but as to the class. What therefore 'introduction' is among the higher classes in Spaceland, that the process of 'feeling' is with us. 'Permit me to ask you to feel and be felt by my friend Mr. So-and-so'—is still, among the more old-fashioned of our country gentlemen in districts remote from towns, the customary formula for a Flatland introduction. But in the towns, and among men of business, the words 'be felt by' are omitted and the sentence is abbreviated to, 'Let me ask you to feel Mr. So-and-so'; although it is assumed, of course, that the 'feeling' is to be reciprocal. Among our still more modern and dashing young gentlemen—who are extremely averse to superfluous effort and supremely indifferent to the purity of their native language—the formula is still further curtailed by the use of 'to feel' in a technical sense, meaning, 'to recommend-for-the-purposes-of-feeling-and-being-felt'; and at this moment the 'slang' of polite or fast society in the upper classes sanctions such a barbarism as 'Mr. Smith, permit me to feel you Mr. Jones.'

Let not my Reader however suppose that 'feeling' is with us the tedious process that it would be with you, or that we find it necessary to feel right round all the sides of every individual before we determine the class to which he belongs. Long practice and training, begun in the schools and continued in the experience of daily life, enable us to discriminate at once by the sense of touch, between the angles of an equal-sided Triangle, Square, and Pentagon; and I need not say that the brainless vertex of an acute-angled Isosceles is obvious to the dullest touch. It is therefore not necessary, as a rule, to do more than feel a single angle of any individual; and this, once ascertained, tells us the class of the person whom we are addressing, unless indeed he belongs to the higher sections of the nobility. There the difficulty is much greater. Even a Master of Arts in our University of Wentbridge* has been known to confuse a ten-sided with a twelve-sided Polygon; and there is hardly a Doctor of Science in or out of that famous University who could pretend to decide promptly and unhesitatingly between a twenty-sided and a twenty-four sided member of the Aristocracy.

Those of my readers who recall the extracts I gave above from the Legislative code concerning Women, will readily perceive that the process of introduction by contact requires some care and discretion. Otherwise the angles might inflict on the unwary Feeler irreparable injury. It is essential for the safety of the Feeler that the Felt should stand perfectly still. A start, a fidgety shifting of the position, yes, even a violent sneeze, has been known before now to prove fatal to the incautious, and to nip in the bud many a promising friendship. Especially is this true among the lower classes of the Triangles. With them, the eye is situated so far from their vertex that they can scarcely take cognizance of what goes on at that extremity of their frame. They are moreover of a rough coarse nature, not sensitive to the delicate touch of the highly organized Polygon. What wonder then if an involuntary toss of the head has ere now deprived the State of a valuable life!

I have heard that my excellent Grandfather—one of the least irregular of his unhappy Isosceles class, who indeed obtained, shortly before his decease, four out of seven votes from the Sanitary and Social Board for passing him into the class of the Equal-sided—often deplored, with a tear in his venerable eye, a miscarriage of this kind, which had occurred to his great-great-great-Grandfather, a respectable Working Man with an angle or brain of 59° 30′. According to his account, my unfortunate Ancestor, being afflicted with rheumatism, and in the act of being felt by a Polygon, by one sudden start accidentally transfixed the Great Man through the diagonal; and thereby, partly in consequence of his long imprisonment and degradation, and partly because of the moral shock which pervaded the whole of my Ancestor's relations, threw back our family a degree and a half in their ascent towards better things. The result was that in the next generation the family brain was registered at only 58°, and not till the lapse of five generations was the lost ground recovered, the full 60° attained, and the Ascent from the Isosceles finally achieved. And all this series of calamities from one little accident in the process of Feeling.

At this point I think I hear some of my better educated readers exclaim, 'How could you in Flatland know anything about angles and degrees, or minutes? We can *see* an angle, because we, in the region of Space, can see two straight lines inclined to one another; but you, who can see nothing but one straight line at a time, or at all events only a number of bits of straight lines all in one straight line,—how can you ever discern any angle, and much less register angles of different sizes?'

I answer that though we cannot *see* angles, we can *infer* them, and this with great precision. Our sense of touch, stimulated by necessity, and developed by long training, enables us to distinguish angles far more accurately than your sense of sight, when unaided by a rule or measure of angles. Nor must I omit to explain that we have great natural helps. It is with us a Law of Nature that the brain of the Isosceles class shall begin at half a degree, or thirty minutes, and shall increase (if it increases at all) by half a degree in every generation; until the goal of 60° is reached, when the condition of serfdom is quitted, and the freeman enters the class of Regulars.

Consequently, Nature herself supplies us with an ascending scale or Alphabet of angles for half a degree up to 60°, specimens of which are placed in every Elementary School throughout the land. Owing to occasional retrogressions, to still more frequent moral and intellectual stagnation, and to the extraordinary fecundity of the Criminal and Vagabond Classes, there is always a vast superfluity of individuals of the half degree and single degree class, and a fair abundance of Specimens up to 10°. These are absolutely destitute of civic rights; and a great number of them, not having even intelligence enough for the purposes of warfare, are devoted by the States to the service of education. Fettered immovably so as to remove all possibility of danger, they are placed in the class rooms of our Infant Schools, and there they are utilized by the Board of Education for the purpose of imparting to the offspring of the Middle Classes that tact and intelligence of which these wretched creatures themselves are utterly devoid.

In some states the Specimens are occasionally fed and suffered to exist for several years; but in the more temperate and better regulated regions, it is found in the long run more advantageous for the educational interests of the young, to dispense with food, and to renew the Specimens every month,—which is about the average duration of the foodless existence of the Criminal class. In the cheaper schools, what is gained by the longer existence of the Specimens is lost, partly in the expenditure for food, and partly in the diminished accuracy of the angles, which are impaired after a few weeks of constant 'feeling.' Nor must we forget to add, in enumerating the advantages of the more expensive system, that it tends, though slightly yet perceptibly, to the diminution of the redundant Isosceles population—an object which every statesman in Flatland constantly keeps in view. On the whole therefore—although I am not ignorant that, in many popularly elected School Boards, there is a reaction in favour of 'the cheap system,'* as it is called—I am myself disposed to think that this is one of the many cases in which expense is the truest economy.

But I must not allow questions of School Board politics to divert me from my subject. Enough has been said, I trust, to show that Recognition by Feeling is not so tedious or indecisive a process as might have been supposed; and it is obviously more trustworthy than Recognition by hearing. Still there remains, as has been pointed out above, the objection that this method is not without danger. For this reason many in the Middle and Lower classes, and all without exception in the Polygonal and Circular orders, prefer a third method, the description of which shall be reserved for the next section.

6

Of Recognition by Sight

I AM about to appear very inconsistent. In previous sections I have said that all figures in Flatland present the appearance of a straight line; and it was added or implied, that it is consequently impossible to distinguish by the visual organ between individuals of different classes: yet now I am about to explain to my Spaceland Critics how we are able to recognize one another by the sense of sight.

If however the Reader will take the trouble to refer to the passage in which Recognition by Feeling is stated to be universal, he will find this qualification—'among the lower classes.' It is only among the higher classes and in our more temperate climates that Sight Recognition is practised.

That this power exists in any regions and for any classes, is the result of Fog; which prevails during the greater part of the year in all parts save the torrid zones. That which is with you in Spaceland an unmixed evil, blotting out the landscape, depressing the spirits, and enfeebling the health, is by us recognized as a blessing scarcely inferior to air itself, and as the Nurse of arts and Parent of sciences. But let me explain my meaning, without further eulogies on this beneficent Element.

If Fog were non-existent, all lines would appear equally and indistinguishably clear; and this is actually the case in those unhappy countries in which the atmosphere is perfectly dry and transparent. But wherever there is a rich supply of Fog, objects that are at a distance, say of three feet, are appreciably dimmer than those at a distance of two feet eleven inches; and the result is that by careful and constant experimental observation of comparative dimness and clearness, we are enabled to infer with great exactness the configuration of the object observed.

An instance will do more than a volume of generalities to make my meaning clear.

Suppose I see two individuals approaching whose rank I wish to ascertain. They are, we will suppose, a Merchant and a Physician, or in other words, an Equilateral Triangle and a Pentagon: how am I to distinguish them?

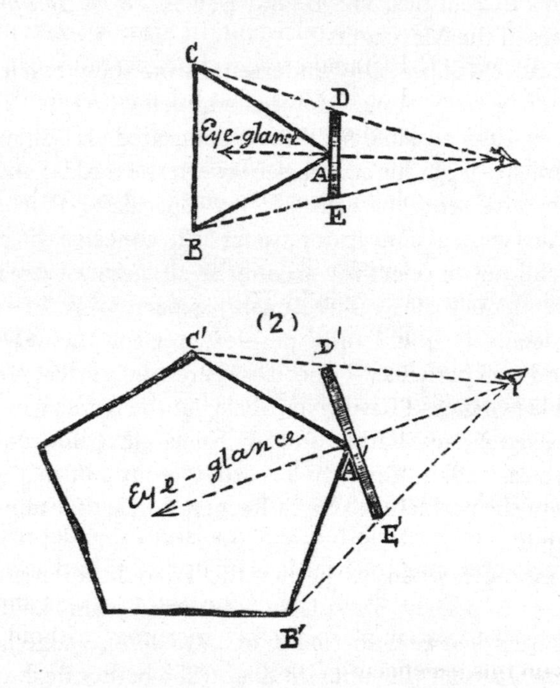

It will be obvious, to every child in Spaceland who has touched the threshold of Geometrical Studies, that, if I can bring my eye so that its glance may bisect an angle (A) of the approaching stranger, my view will lie as it were evenly between his two sides that are next to me (viz. CA and AB), so that I shall contemplate the two impartially, and both will appear of the same size.

Now in the case of (1) the Merchant, what shall I see? I shall see a straight line DAE, in which the middle point (A) will be very bright because it is nearest to me; but on either side the line will shade away *rapidly into dimness*, because the sides AC and AB *recede rapidly into the fog*; and what appear to me as the Merchant's extremities, viz. D and E, will be *very dim indeed*.

On the other hand in the case of (2) the Physician, though I shall here also see a line (D′A′E′) with a bright centre (A′), yet it will shade away *less rapidly* into dimness, because the sides (A′C′, A′B′) *recede less rapidly into the fog*; and what appear to me the Physician's extremities, viz. D′ and E′, will be *not so dim* as the extremities of the Merchant.

The Reader will probably understand from these two instances how—after a very long training supplemented by constant experience—it is possible for the well-educated classes among us to discriminate with fair accuracy between the middle and lowest orders, by the sense of sight. If my Spaceland Patrons have grasped this general conception, so far as to conceive the possibility of it and not to reject my account as altogether incredible—I shall have attained all I can reasonably expect. Were I to attempt further details I should only perplex. Yet for the sake of the young and inexperienced, who may perchance infer—from the two simple instances I have given above, of the manner in which I should recognize my Father and my Sons—that Recognition by sight is an easy affair, it may be needful to point out that in actual life most of the problems of Sight Recognition are far more subtle and complex.

If for example, when my Father, the Triangle, approaches me, he happens to present his side to me instead of his angle, then, until I have asked him to rotate, or until I have edged my eye round him, I am for the moment doubtful whether he may not be a Straight Line, or, in other words, a Woman. Again, when I am in the company of one of my two hexagonal Grandsons, contemplating one of his sides (AB) full front, it will be evident from the accompanying diagram that I shall see one whole line (AB) in comparative brightness (shading off hardly at all at the ends) and two smaller lines (CA and BD) dim throughout and shading away into greater dimness toward the extremities C and D.

But I must not give way to the temptation of enlarging on these topics. The meanest mathematician in Spaceland will readily believe me when I assert that the problems of life, which present

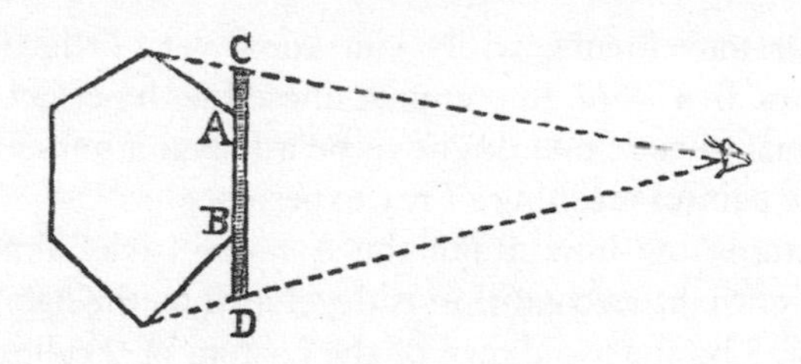

themselves to the well-educated—when they are themselves in motion, rotating, advancing or retreating, and at the same time attempting to discriminate by the sense of sight between a number of Polygons of high rank moving in different directions, as for example in a ball-room or conversazione—must be of a nature to task the angularity of the most intellectual, and amply justify the rich endowments of the Learned Professors of Geometry, both Static and Kinetic, in the illustrious University of Wentbridge, where the Science and Art of Sight Recognition are regularly taught to large classes of the *élite* of the States.

It is only a few of the scions of our noblest and wealthiest houses, who are able to give thc time and money necessary for the thorough prosecution of this noble and valuable Art. Even to me, a Mathematician of no mean standing, and the Grandfather of two most hopeful and perfectly regular Hexagons, to find myself in the midst of a crowd of rotating Polygons of the higher classes, is occasionally very perplexing. And of course to a common Tradesman, or Serf, such a sight is almost as unintelligible as it would be to you, my Reader, were you suddenly transported into our country.

In such a crowd you could see on all sides of you nothing but a Line, apparently straight, but of which the parts would vary irregularly and perpetually in brightness or dimness. Even if you had completed your third year in the Pentagonal and Hexagonal classes in the University, and were perfect in the theory of the subject, you would still find that there was need of many years of experience, before you could move in a fashionable crowd without jostling against your betters, whom it is against etiquette to ask to 'feel,' and who, by their superior culture and breeding, know all about your movements, while you know very little or nothing about theirs. In a word, to comport oneself with perfect propriety in Polygonal society, one ought to be a Polygon oneself. Such at least is the painful teaching of my experience.

It is astonishing how much the Art—or I may almost call it instinct—of Sight Recognition is developed by the habitual practice of it and by the avoidance of the custom of 'Feeling.' Just as, with you, the deaf and dumb, if once allowed to gesticulate and to use the hand-alphabet, will never acquire the more difficult but far more valuable art of lip-speech and lip-reading, so it is with us as regards 'Seeing' and 'Feeling.' None who in early life resort to 'Feeling' will ever learn 'Seeing' in perfection.

For this reason, among our Higher Classes, 'Feeling' is discouraged or absolutely forbidden. From the cradle their children, instead of going to the Public Elementary schools (where the art of Feeling is taught,) are sent to higher Seminaries of an exclusive character; and at our illustrious University, to 'feel' is regarded as a most serious fault, involving Rustication* for the first offence, and Expulsion for the second.

But among the lower classes the art of Sight Recognition is regarded as an unattainable luxury. A common Tradesman cannot afford to let his son spend a third of his life in abstract studies.* The children of the poor are therefore allowed to 'feel' from their earliest years, and they gain thereby a precocity and an early vivacity which contrast at first most favourably with the inert, undeveloped, and listless behaviour of the half-instructed youths of the Polygonal class; but when the latter have at last completed their University course, and are prepared to put their theory into practice, the change that comes over them may almost be described as a new birth, and in every art, science, and social pursuit they rapidly overtake and distance their Triangular competitors.

Only a few of the Polygonal Class fail to pass the Final Test or Leaving Examination at the University. The condition of the unsuccessful minority is truly pitiable. Rejected from the higher class, they are also despised by the lower. They have neither the matured and systematically trained powers of the Polygonal

Bachelors and Masters of Arts, nor yet the native precocity and mercurial versatility of the youthful Tradesman. The professions, the public services are closed against them; and though in most States they are not actually debarred from marriage, yet they have the greatest difficulty in forming suitable alliances, as experience shows that the offspring of such unfortunate and ill-endowed parents is generally itself unfortunate, if not positively Irregular.

It is from these specimens of the refuse of our Nobility that the great Tumults and Seditions of past ages have generally derived their leaders; and so great is the mischief thence arising that an increasing minority of our more progressive Statesmen are of opinion that true mercy would dictate their entire suppression, by enacting that all who fail to pass the Final Examination of the University should be either imprisoned for life, or extinguished by a painless death.

But I find myself digressing into the subject of Irregularities, a matter of such vital interest that it demands a separate section.

7
*Of Irregular Figures**

THROUGHOUT the previous pages I have been assuming—what perhaps should have been laid down at the beginning as a distinct and fundamental proposition—that every human being in Flatland is a Regular Figure, that is to say of regular construction. By this I mean that a Woman must not only be a line, but a straight line; that an Artisan or Soldier must have two of his sides equal; that Tradesmen must have three sides equal; Lawyers (of which class I am a humble member), four sides equal, and, generally, that in every Polygon, all the sides must be equal.

The size of the sides would of course depend upon the age of the individual. A Female at birth would be about an inch long, while a tall adult Woman might extend to a foot. As to the Males of every class, it may be roughly said that the length of an adult's sides, when added together, is two feet or a little more. But the size of our sides is not under consideration. I am speaking of the *equality* of sides, and it does not need much reflection to see that the whole of the social life in Flatland rests upon the fundamental fact that Nature wills all Figures to have their sides equal.

If our sides were unequal our angles might be unequal. Instead of its being sufficient to feel, or estimate by sight, a single angle in order to determine the form of an individual, it would be necessary to ascertain each angle by the experiment of Feeling. But life would be too short for such a tedious groping. The whole science and art of Sight Recognition would at once perish; Feeling, so far as it is an art, would not long survive; intercourse would become perilous or impossible; there would be an end to all confidence, all forethought; no one would be safe in making the most simple social arrangements; in a word, civilization would relapse into barbarism.

Am I going too fast to carry my Readers with me to these

obvious conclusions? Surely a moment's reflection, and a single instance from common life, must convince every one that our whole social system is based upon Regularity, or Equality of Angles. You meet, for example, two or three Tradesmen in the street, whom you recognize at once to be Tradesmen by a glance at their angles and rapidly bedimmed sides, and you ask them to step into your house to lunch. This you do at present with perfect confidence, because every one knows to an inch or two the area occupied by an adult Triangle: but imagine that your Tradesman drags behind his regular and respectable vertex, a parallelogram of twelve or thirteen inches in diagonal:—what are you to do with such a monster sticking fast in your house door?

But I am insulting the intelligence of my Readers by accumulating details which must be patent to every one who enjoys the advantages of a Residence in Spaceland. Obviously the measurements of a single angle would no longer be sufficient under such portentous circumstances; one's whole life would be taken up in feeling or surveying the perimeter of one's acquaintances. Already the difficulties of avoiding a collision in a crowd are enough to tax the sagacity of even a well-educated Square; but if no one could calculate the Regularity of a single figure in the company, all would be chaos and confusion, and the slightest panic would cause serious injuries, or—if there happened to be any women or Soldiers present—perhaps considerable loss of life.

Expediency therefore concurs with Nature in stamping the seal of its approval upon Regularity of conformation: nor has the Law been backward in seconding their efforts. 'Irregularity of Figure' means with us the same as, or more than, a combination of moral obliquity and criminality with you, and is treated accordingly. There are not wanting, it is true, some promulgators of paradoxes who maintain that there is no necessary connection between geometrical and moral Irregularity. 'The Irregular,' they say, 'is from his birth scouted by his own parents, derided by his brothers and sisters, neglected by the domestics, scorned and suspected by society, and excluded from all posts of responsibility, trust, and useful activity. His every movement is jealously watched by the police till he comes of age and presents himself for inspection; then he is either destroyed, if he is found to exceed the fixed margin of deviation, or else immured in a Government Office as a clerk of the seventh class; prevented from marriage; forced to drudge at an uninteresting occupation for a miserable stipend; obliged to live and board at the office, and to take even his vacation under close supervision; what wonder that human nature, even in the best and purest, is embittered and perverted by such surroundings!'

All this very plausible reasoning does not convince me, as it has not convinced the wisest of our Statesmen, that our ancestors erred in laying it down as an axiom of policy that the toleration of Irregularity is incompatible with the safety of the State. Doubtless, the life of an Irregular is hard; but the interests of the Greater Number require that it shall be hard. If a man with a triangular front and a polygonal back were allowed to exist and to propagate a still more Irregular posterity, what would become of the arts of life? Are the houses and doors and churches in Flatland to be altered in order to accommodate such monsters? Are our ticket-collectors to be required to measure every man's perimeter before they allow him to enter a theatre, or to take his place in a lecture room? Is an Irregular to be exempted from the militia? And if not, how is he to be prevented from carrying desolation into the ranks of his comrades? Again, what irresistible temptations to fraudulent impostures must needs beset such a creature! How easy for him to enter a shop with his polygonal front foremost, and to order goods to any extent from a confiding Tradesman! Let the advocates of a falsely called Philanthropy plead as they may for the abrogation of the Irregular Penal Laws, I for my part have never known an Irregular who was not also what Nature evidently intended him to be—a hypocrite, a misanthropist, and, up to the limits of his power—a perpetrator of all manner of mischief.

Not that I should be disposed to recommend (at present) the extreme measures adopted in some States, where an infant whose angle deviates by half a degree from the correct angularity is summarily destroyed at birth. Some of our highest and ablest men, men of real genius, have during their earliest days laboured under deviations as great as, or even greater than, forty-five minutes: and the loss of their precious lives would have been an irreparable injury to the State. The art of healing also has achieved some of its most glorious triumphs in the compressions, extensions, trepannings, colligations, and other surgical or diætetic operations by which Irregularity has been partly or wholly cured. Advocating therefore a *Via Media*,* I would lay down no fixed or absolute line of demarcation; but at the period when the frame is just beginning to set, and when the Medical Board has reported that recovery is improbable, I would suggest that the Irregular offspring be painlessly and mercifully consumed.

8

Of the Ancient Practice of Painting

If my Readers have followed me with any attention up to this point, they will not be surprised to hear that life is somewhat dull in Flatland. I do not, of course, mean that there are not battles, conspiracies, tumults, factions, and all those other phenomena which are supposed to make History interesting; nor would I deny that the strange mixture of the problems of life and the problems of Mathematics, continually inducing conjecture and giving the opportunity of immediate verification, imparts to our existence a zest which you in Spaceland can hardly comprehend. I speak now from the æsthetic and artistic point of view when I say that life with us is dull; æsthetically and artistically, very dull indeed.

How can it be otherwise, when all one's prospect, all one's landscapes, historical pieces, portraits, flowers, still life, are nothing but a single line, with no varieties except degrees of brightness and obscurity?

It was not always thus. Colour, if Tradition speaks the truth, once for the space of half a dozen centuries or more, threw a transient splendour over the lives of our ancestors in remote ages. Some private individual—a Pentagon whose name is variously reported—having casually discovered the constituents of the simpler colours and a rudimentary method of painting, is said to have begun by decorating first his house, then his slaves, then his Father, his Sons and Grandsons, lastly himself. The convenience as well as the beauty of the results commended themselves to all. Wherever Chromatistes*—for by that name the most trustworthy authorities concur in calling him,—turned his variegated frame, there he at once excited attention, and attracted respect. No one now needed to 'feel' him; no one mistook his front for his back; all his movements were readily ascertained by his neighbours without the slightest strain on their powers of calculation; no one jostled him, or failed to make way for him; his voice was saved the labour of that exhausting utterance by which we colourless Squares and Pentagons are often forced to proclaim our individuality when we move amid a crowd of ignorant Isosceles.

The fashion spread like wildfire. Before a week was over, every Square and Triangle in the district had copied the example of Chromatistes, and only a few of the more conservative Pentagons still held out. A month or two found even the Dodecagons infected with the innovation. A year had not elapsed before the habit had spread to all but the very highest of the Nobility. Needless to say, the custom soon made its way from the district of Chromatistes to surrounding regions; and within two generations no one in all Flatland was colourless except the Women and the Priests.

Here Nature herself appeared to erect a barrier, and to plead against extending the innovation to these two classes. Many-sidedness was almost essential as a pretext for the Innovators. 'Distinction of sides is intended by Nature to imply distinction of colours'—such was the sophism which in those days flew from mouth to mouth, converting whole towns at a time to the new culture. But manifestly to our Priests and Women this adage did not apply. The latter had only one side, and therefore—plurally and pedantically speaking—*no sides*. The former—if at least they would assert their claim to be really and truly Circles, and not mere high-class Polygons with an infinitely large number of infinitesimally small sides—were in the habit of boasting (what Women confessed and deplored) that they also had no sides, being blessed with a perimeter of one line or, in other words, a Circumference. Hence it came to pass that these two Classes could see no force in the so-called axiom about 'Distinction of Sides implying Distinction of Colour'; and when all others had succumbed to the fascinations of corporal decoration, the Priests and the Women alone still remained pure from the pollution of paint.

Immoral, licentious, anarchical, unscientific—call them by what names you will—yet, from an æsthetic point of view, those ancient days of the Colour Revolt were the glorious childhood of Art in Flatland—a childhood, alas, that never ripened into manhood, nor even reached the blossom of youth. To live was then in itself a delight, because living implied seeing. Even at a small party, the company was a pleasure to behold; the richly varied hues of the assembly in a church or theatre are said to have more than once proved too distracting for our greatest teachers and actors; but most ravishing of all is said to have been the unspeakable magnificence of a military review.

The sight of a line of battle of twenty thousand Isosceles suddenly facing about, and exchanging the sombre black of their bases for the orange and purple of the two sides including their acute angle; the militia of the Equilateral Triangles tricoloured in red, white, and blue; the mauve, ultramarine, gamboge, and burnt umber of the Square artillerymen rapidly rotating near their vermilion guns; the dashing and flashing of the five-coloured and six-coloured Pentagons and Hexagons careering across the field in their offices of surgeons, geometricians and aides-de-camp—all these may well have been sufficient to render credible the famous story how an illustrious Circle, overcome by the artistic beauty of the forces under his command, threw aside his marshal's bâton and his royal crown, exclaiming that he henceforth exchanged them for the artist's pencil. How great and glorious the sensuous development of these days must have been is in part indicated by the very language and vocabulary of the period.* The commonest utterances of the commonest citizens in the time of the Colour Revolt seem to have been suffused with a richer tinge of word or thought; and to that era we are even now indebted for our finest poetry and for whatever rhythm still remains in the more scientific utterance of these modern days.

9

*Of the Universal Colour Bill**

BUT meanwhile the intellectual Arts were fast decaying.

The Art of Sight Recognition, being no longer needed, was no longer practised; and the studies of Geometry, Statics, Kinetics, and other kindred subjects, came soon to be considered superfluous, and fell into disrepute and neglect even at our University. The inferior Art of Feeling speedily experienced the same fate at our Elementary Schools. Then the Isosceles classes, asserting that the Specimens were no longer used nor needed, and refusing to pay the customary tribute from the Criminal classes to the service of Education, waxed daily more numerous and more insolent on the strength of their immunity from the old burden which had formerly exercised the twofold wholesome effect of at once taming their brutal nature and thinning their excessive numbers.

Year by year the Soldiers and Artisans began more vehemently to assert—and with increasing truth—that there was no great difference between them and the very highest class of Polygons, now that they were raised to an equality with the latter, and enabled to grapple with all the difficulties and solve all the problems of life, whether Statical and Kinetical, by the simple process of Colour Recognition. Not content with the natural neglect into which Sight Recognition was falling, they began boldly to demand the legal prohibition of all 'monopolising and aristocratic Arts' and the consequent abolition of all endowments for the studies of Sight Recognition, Mathematics, and Feeling. Soon, they began to insist that inasmuch as Colour, which was a second Nature, had destroyed the need of aristocratic distinctions, the Law should follow in the same path, and that henceforth all individuals and all classes should be recognized as absolutely equal and entitled to equal rights.

Finding the higher Orders wavering and undecided, the leaders of the Revolution advanced still further in their requirements, and at last demanded that all classes alike, the Priests and the Women not excepted, should do homage to Colour by submitting to be painted. When it was objected that Priests and Women had no sides, they retorted that Nature and Expediency concurred in dictating that the front half of every human being (that is to say, the half containing his eye and mouth) should be distinguishable from his hinder half. They therefore brought before a general and extraordinary Assembly of all the States of Flatland a Bill proposing that in every Woman the half containing the eye and mouth should be coloured red, and the other half green. The Priests were to be painted in the same way, red being applied to that semicircle in which the eye and mouth formed the middle point; while the other or hinder semicircle was to be coloured green.

There was no little cunning in this proposal, which indeed emanated, not from any Isosceles—for no being so degraded would have had angularity enough to appreciate, much less to devise, such a model of state-craft—but from an Irregular Circle who, instead of being destroyed in his childhood, was reserved by a foolish indulgence to bring desolation on his country and destruction on myriads of his followers.

On the one hand the proposition was calculated to bring the Women in all classes over to the side of the Chromatic Innovation. For by assigning to the Women the same two colours as were assigned to the Priests, the Revolutionists thereby ensured that, in certain positions, every Woman would appear like a Priest, and be treated with corresponding respect and deference—a prospect that could not fail to attract the Female Sex in a mass.

But by some of my Readers the possibility of the identical appearance of Priests and Women, under the new Legislation, may not be recognized; if so, a word or two will make it obvious.

Imagine a woman duly decorated, according to the new Code; with the front half (*i.e.* the half containing eye and mouth) red, and with the hinder half green. Look at her from one side. Obviously you will see a straight line, *half red, half green*.

Now imagine a Priest, whose mouth is at M, and whose front semicircle (AMB) is consequently coloured red, while his hinder semicircle is green; so that the diameter AB divides the green from the red. If you contemplate the Great Man so as to have your eye in the same straight line as his dividing diameter (AB), what you will see will be a straight line (CBD), of which *one half* (CB) *will be red, and the other* (BD) *green*. The whole line (CD) will be rather shorter perhaps than that of a full-sized Woman, and will shade off more rapidly towards its extremities; but the identity of the colours would give you an immediate impression of identity of Class, making you neglectful of other details. Bear in mind the decay of Sight Recognition which threatened society at the time of the Colour Revolt; add too the certainty that Women would speedily learn to shade off their extremities so as to imitate the Circles; it must then be surely obvious to you, my dear Reader, that the Colour Bill placed us under a great danger of confounding a Priest with a young Woman.

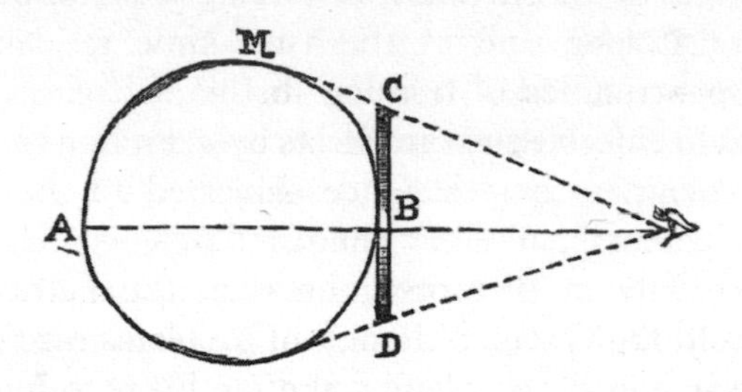

How attractive this prospect must have been to the Frail Sex may readily be imagined. They anticipated with delight the confusion that would ensue. At home they might hear political and ecclesiastical secrets intended not for them but for their husbands and brothers, and might even issue commands in the name of a priestly Circle; out of doors the striking combination of red and green, without addition of any other colours, would be sure to lead the common people into endless mistakes, and the Women would gain whatever the Circles lost, in the deference of the passers by. As for the scandal that would befall the Circular Class if the frivolous and unseemly conduct of the Women were imputed to them, and as to the consequent subversion of the Constitution, the Female Sex could not be expected to give a thought to these considerations. Even in the households of the Circles, the Women were all in favour of the Universal Colour Bill.

The second object aimed at by the Bill was the gradual demoralization of the Circles themselves. In the general intellectual decay they still preserved their pristine clearness and strength of understanding. From their earliest childhood, familiarized in their Circular households with the total absence of Colour, the Nobles alone preserved the Sacred Art of Sight Recognition, with all the advantages that result from that admirable training of the intellect. Hence, up to the date of the introduction of the Universal Colour Bill, the Circles had not only held their own, but even increased their lead of other classes by abstinence from the popular fashion.

Now therefore the artful Irregular whom I described above as the real author of this diabolical Bill, determined at one blow to lower the status of the Hierarchy by forcing them to submit to the pollution of Colour, and at the same time to destroy their domestic opportunities of training in the Art of Sight Recognition, so as to enfeeble their intellects by depriving them of their pure and colourless homes. Once subjected to the chromatic taint, every parental and every childish Circle would demoralize each other. Only in discerning between the Father and the Mother would the Circular infant find problems for the exercise of its understanding—problems too often likely to be corrupted by maternal impostures with the result of shaking the child's faith in all logical conclusions. Thus by degrees the intellectual lustre of the Priestly Order would wane, and the road would then lie open for a total destruction of all Aristocratic Legislature and for the subversion of our Privileged Classes.

10

Of the Suppression of the Chromatic Sedition

THE agitation for the Universal Colour Bill continued for three years; and up to the last moment of that period it seemed as though Anarchy were destined to triumph.

A whole army of Polygons, who turned out to fight as private soldiers, was utterly annihilated by a superior force of Isosceles Triangles—the Squares and Pentagons meanwhile remaining neutral. Worse than all, some of the ablest Circles fell a prey to conjugal fury. Infuriated by political animosity, the wives in many a noble household wearied their lords with prayers to give up their opposition to the Colour Bill; and some, finding their entreaties fruitless, fell on and slaughtered their innocent children and husbands, perishing themselves in the act of carnage. It is recorded that during that triennial agitation no less than twenty-three Circles perished in domestic discord.

Great indeed was the peril. It seemed as though the Priests had no choice between submission and extermination; when suddenly the course of events was completely changed by one of those picturesque incidents which Statesmen ought never to neglect, often to anticipate, and sometimes perhaps to originate, because of the absurdly disproportionate power with which they appeal to the sympathies of the populace.

It happened that an Isosceles of a low type, with a brain little if at all above four degrees—accidentally dabbling in the colours of some Tradesman whose shop he had plundered—painted himself, or caused himself to be painted (for the story varies) with the twelve colours of a Dodecagon. Going into the Market Place he accosted in a feigned voice a maiden, the orphan daughter of a noble Polygon, whose affection in former days he had sought in vain; and by a series of deceptions—aided, on the one side, by a string of lucky accidents too long to relate, and, on the other, by an almost inconceivable fatuity and neglect of ordinary precautions on the part of the relations of the bride—he succeeded in consummating the marriage. The unhappy girl committed suicide on discovering the fraud to which she had been subjected.

When the news of this catastrophe spread from State to State the minds of the Women were violently agitated. Sympathy with the miserable victim and anticipations of similar deceptions for themselves, their sisters, and their daughters, made them now regard the Colour Bill in an entirely new aspect. Not a few openly avowed themselves converted to antagonism; the rest needed only a slight stimulus to make a similar avowal. Seizing this favourable opportunity, the Circles hastily convened an extraordinary Assembly of the States; and besides the usual guard of Convicts, they secured the attendance of a large number of reactionary Women.

Amidst an unprecedented concourse, the Chief Circle of those days—by name Pantocyclus*—arose to find himself hissed and hooted by a hundred and twenty thousand Isosceles. But he secured silence by declaring that henceforth the Circles would enter on a policy of Concession; yielding to the wishes of the majority, they would accept the Colour Bill. The uproar being at once converted to applause, he invited Chromatistes, the leader of the Sedition, into the centre of the hall, to receive in the name of his followers the submission of the Hierarchy. Then followed a speech, a masterpiece of rhetoric, which occupied nearly a day in the delivery, and to which no summary can do justice.

With a grave appearance of impartiality he declared that, as they were now finally committing themselves to Reform or Innovation, it was desirable that they should take one last view of the perimeter of the whole subject, its defects as well as its advantages. Gradually introducing the mention of the dangers to the Tradesmen, the Professional Classes and the Gentlemen, he silenced the rising murmurs of the Isosceles by reminding them that, in spite of all these defects, he was willing to accept the Bill if it was approved by the majority. But it was manifest that all, except the Isosceles, were moved by his words and were either neutral or averse to the Bill.

Turning now to the Workmen he asserted that their interests must not be neglected, and that, if they intended to accept the Colour Bill, they ought at least to do so with a full view of the consequences. Many of them, he said, were on the point of being admitted to the class of the Regular Triangles; others anticipated for their children a distinction they could not hope for themselves. That honourable ambition would now have to be sacrificed. With the universal adoption of Colour, all distinctions would cease; Regularity would be confused with Irregularity; development would give place to retrogression; the Workman would in a few generations be degraded to the level of the Military, or even the Convict Class; political power would be in the hands of the greatest number, that is to say the Criminal Classes; who were already more numerous than the Workmen, and would soon out-number all the other Classes put together when the usual Compensative Laws of Nature were violated.

A subdued murmur of assent ran through the ranks of the Artisans, and Chromatistes, in alarm, attempted to step forward and address them. But he found himself encompassed with guards and forced to remain silent while the Chief Circle in a few impassioned words made a final appeal to the Women, exclaiming that, if the Colour Bill passed, no marriage would henceforth be safe, no woman's honour secure; fraud, deception, hypocrisy would pervade every household; domestic bliss would share the fate of the Constitution and pass to speedy perdition: 'Sooner than this,' he cried, 'Come death.'

At these words, which were the preconcerted signal for action, the Isosceles Convicts fell on and transfixed the wretched Chromatistes; the Regular Classes, opening their ranks, made way for a band of Women who, under direction of the Circles, moved, back foremost, invisibly and unerringly upon the unconscious Soldiers; the Artisans, imitating the example of their betters, also opened their ranks. Meantime bands of Convicts occupied every entrance with an impenetrable phalanx.

The battle, or rather carnage, was of short duration. Under the skilful generalship of the Circles almost every Woman's charge was fatal, and very many extracted their sting uninjured, ready for a second slaughter. But no second blow was needed; the rabble of the Isosceles did the rest of the business for themselves. Surprised, leader-less, attacked in front by invisible foes, and finding egress cut off by the Convicts behind them, they at once—after their manner—lost all presence of mind, and raised the cry of 'treachery.' This sealed their fate. Every Isosceles now saw and felt a foe in every other. In half an hour not one of that vast multitude was living; and the fragments of seven score thousand of the Criminal Class slain by one another's angles attested the triumph of Order.

The Circles delayed not to push their victory to the uttermost. The Working Men they spared but decimated. The Militia of the Equilaterals was at once called out; and every Triangle suspected of Irregularity on reasonable grounds, was destroyed by Court Martial, without the formality of exact measurement by the Social Board. The homes of the Military and Artisan classes were inspected in a course of visitations extending through upwards of a year; and during that period every town, village, and hamlet was systematically purged of that excess of the lower orders which had been brought about by the neglect to pay the Tribute of Criminals to the Schools and University, and by the violation of the other natural Laws of the Constitution of Flatland. Thus the balance of classes was again restored.

Needless to say that henceforth the use of Colour was abolished, and its possession prohibited. Even the utterance of any word denoting Colour, except by the Circles or by qualified scientific teachers, was punished by a severe penalty. Only at our University in some of the very highest and most esoteric classes—which I myself have never been privileged to attend—it is understood that the sparing use of Colour is still sanctioned for the purpose of illustrating some of the deeper problems of mathematics. But of this I can only speak from hearsay.

Elsewhere in Flatland, Colour is now non-existent. The art of making it is known to only one living person, the Chief Circle for the time being; and by him it is handed down on his death-bed to none but his Successor. One manufactory alone produces it; and, lest the secret should be betrayed, the Workmen are annually consumed, and fresh ones introduced. So great is the terror with which even now our Aristocracy looks back to the far-distant days of the agitation for the Universal Colour Bill.

II

Concerning our Priests

IT is high time that I should pass from these brief and discursive notes about things in Flatland to the central event of this book, my initiation into the mysteries of Space. *That* is my subject; all that has gone before is merely preface.

For this reason I must omit many matters of which the explanation would not, I flatter myself, be without interest for my Readers: as for example, our method of propelling and stopping ourselves, although destitute of feet; the means by which we give fixity to structures of wood, stone, or brick, although of course we have no hands, nor can we lay foundations as you can, nor avail ourselves of the lateral pressure of the earth; the manner in which the rain originates in the intervals between our various zones, so that the northern regions do not intercept the moisture from falling on the southern; the nature of our hills and mines, our trees and vegetables, our seasons and harvests; our Alphabet, suited to our linear tablets; our eyes, adapted to our linear sides; these and a hundred other details of our physical existence I must pass over; nor do I mention them now except to indicate to my readers that their omission proceeds, not from forgetfulness on the part of the Author, but from his regard for the time of the Reader.

Yet before I proceed to my legitimate subject some few final remarks will no doubt be expected by my Readers upon those pillars and mainstays of the Constitution of Flatland, the controllers of our conduct and shapers of our destiny, the objects of universal homage and almost of adoration: need I say that I mean our Circles or Priests?

When I call them Priests, let me not be understood as meaning no more than the term denotes with you. With us, our Priests are Administrators of all Business, Art, and Science; Directors of

Trade, Commerce, Generalship, Architecture, Engineering, Education, Statesmanship, Legislature, Morality, Theology; doing nothing themselves, they are the Causes of everything, worth doing, that is done by others.

Although popularly every one called a Circle is deemed a Circle, yet among the better educated Classes it is known that no Circle is really a Circle, but only a Polygon with a very large number of very small sides. As the number of the sides increases, a Polygon approximates to a Circle; and, when the number is very great indeed, say for example three or four hundred, it is extremely difficult for the most delicate touch to feel any polygonal angles. Let me say rather, it *would* be difficult: for, as I have shown above, Recognition by Feeling is unknown among the highest society, and to *feel* a Circle would be considered a most audacious insult. This habit of abstention from Feeling in the best society enables a Circle the more easily to sustain the veil of mystery in which, from his earliest years, he is wont to enwrap the exact nature of his Perimeter or Circumference. Three feet being the average Perimeter, it follows that, in a Polygon of three hundred sides, each side will be no more than the hundredth part of a foot in length, or little more than the tenth part of an inch; and in a Polygon of six or seven hundred sides the sides are little larger than the diameter of a Spaceland pin-head. It is always assumed, by courtesy, that the Chief Circle for the time being has ten thousand sides.

The ascent of the posterity of the Circles in the social scale is not restricted, as it is among the lower Regular classes, by the Law of Nature which limits the increase of sides to one in each generation. If it were so, the number of sides in a Circle would be a mere question of pedigree and arithmetic; and the four hundred and ninety-seventh descendant of an Equilateral Triangle would necessarily be a Polygon with five hundred sides. But this is not the case. Nature's Law prescribes two antagonistic decrees affecting Circular propagation; first, that as the race climbs higher in the scale of development, so development shall proceed at an accelerated pace; second, that in the same proportion, the race shall become less fertile.* Consequently in the home of a Polygon of four or five hundred sides it is rare to find a son; more than one is never seen. On the other hand the son of a five-hundred-sided Polygon has been known to possess five hundred and fifty, or even six hundred sides.

Art also steps in to help the process of the higher Evolution. Our physicians have discovered that the small and tender sides of an infant Polygon of the higher class can be fractured, and his whole frame re-set, with such exactness that a Polygon of two or three hundred sides sometimes—by no means always, for the process is attended with serious risk—but sometimes overleaps two or three hundred generations, and as it were doubles at a stroke, the number of his progenitors and the nobility of his descent.

Many a promising child is sacrificed in this way. Scarcely one out of ten survives. Yet so strong is the parental ambition among those Polygons who are, as it were, on the fringe of the Circular class, that it is very rare to find a Nobleman, of that position in society, who has neglected to place his first-born son in the Circular Neo-Therapeutic Gymnasium before he has attained the age of a month.

One year determines success or failure. At the end of that time the child has, in all probability, added one more to the tombstones that crowd the Neo-Therapeutic Cemetery; but on rare occasions a glad procession bears back the little one to his exultant parents, no longer a Polygon, but a Circle, at least by courtesy: and a single instance of so blessed a result induces multitudes of Polygonal parents to submit to similar domestic sacrifices, which have a dissimilar issue.

12

Of the Doctrine of our Priests

As to the doctrine of the Circles it may briefly be summed up in a single maxim, 'Attend to your Configuration.' Whether political, ecclesiastical, or moral, all their teaching has for its object the improvement of individual and collective Configuration—with special reference of course to the Configuration of the Circles, to which all other objects are subordinated.

It is the merit of the Circles that they have effectually suppressed those ancient heresies which led men to waste energy and sympathy in the vain belief that conduct depends upon will, effort, training, encouragement, praise, or anything else but Configuration.* It was Pantocyclus—the illustrious Circle mentioned above, as the queller of the Colour Revolt—who first convinced mankind that Configuration makes the man; that if, for example, you are born an Isosceles with two uneven sides, you will assuredly go wrong unless you have them made even—for which purpose you must go to the Isosceles Hospital; similarly, if you are a Triangle, or Square, or even a Polygon, born with any Irregularity, you must be taken to one of the Regular Hospitals to have your disease cured; otherwise you will end your days in the State Prison or by the angle of the State Executioner.

All faults or defects, from the slightest misconduct to the most flagitious crime, Pantocyclus attributed to some deviation from perfect Regularity in the bodily figure, caused perhaps (if not congenital) by some collision in a crowd; by neglect to take exercise, or by taking too much of it; or even by a sudden change of temperature, resulting in a shrinkage or expansion in some too susceptible part of the frame. Therefore, concluded that illustrious Philosopher, neither good conduct nor bad conduct is a fit subject, in any sober estimation, for either praise or blame. For why should you praise, for example, the integrity of a Square who faithfully defends the interests of his client, when you ought in reality rather to admire the exact precision of his right angles? Or again, why blame a lying, thievish Isosceles when you ought rather to deplore the incurable inequality of his sides?

Theoretically, this doctrine is unquestionable; but it has practical drawbacks. In dealing with an Isosceles, if a rascal pleads that he cannot help stealing because of his unevenness, you reply that for that very reason, because he cannot help being a nuisance to his neighbours, you, the Magistrate, cannot help sentencing him to be consumed—and there's an end of the matter. But in little domestic difficulties, where the penalty of consumption, or death, is out of the question, this theory of Configuration sometimes comes in awkwardly; and I must confess that occasionally when one of my own Hexagonal Grandsons pleads as an excuse for his disobedience that a sudden change of the temperature has been too much for his Perimeter, and that I ought to lay the blame not on him but on his Configuration, which can only be strengthened by abundance of the choicest sweetmeats, I neither see my way logically to reject, nor practically to accept, his conclusions.

For my own part, I find it best to assume that a good sound scolding or castigation has some latent and strengthening influence on my Grandson's Configuration; though I own that I have no grounds for thinking so. At all events I am not alone in my way of extricating myself from this dilemma; for I find that many of the highest Circles, sitting as Judges in Law courts, use praise and blame towards Regular and Irregular Figures; and in their homes I know by experience that, when scolding their children, they speak about 'right' or 'wrong' as vehemently and passionately as if they believed that these names represented real existences, and that a human Figure is really capable of choosing between them.

Consistently carrying out their policy of making Configuration the leading idea in every mind, the Circles reverse the nature of that Commandment which in Spaceland regulates the relations between parents and children. With you, children are taught to honour their parents; with us—next to the Circles, who are the chief object of universal homage—a man is taught to honour his Grandson, if he has one; or, if not, his Son. By 'honour,' however, is by no means meant 'indulgence,' but a reverent regard for their highest interests: and the Circles teach that the duty of fathers is to subordinate their own interests to those of posterity, thereby advancing the welfare of the whole State as well as that of their own immediate descendants.

The weak point in the system of the Circles—if a humble Square may venture to speak of anything Circular as containing any element of weakness—appears to me to be found in their relations with Women.

As it is of the utmost importance for Society that Irregular births should be discouraged, it follows that no Woman who has any Irregularities in her ancestry is a fit partner for one who desires that his posterity should rise by regular degrees in the social scale.

Now the Irregularity of a Male is a matter of measurement; but as all Women are straight, and therefore visibly Regular, so to speak, one has to devise some other means of ascertaining what I may call their invisible Irregularity, that is to say their potential Irregularities as regards possible offspring. This is effected by carefully-kept pedigrees, which are preserved and supervised by the State; and without a certified pedigree no Woman is allowed to marry.

Now it might have been supposed that a Circle—proud of his ancestry and regardful for a posterity which might possibly issue hereafter in a Chief Circle—would be more careful than any other to choose a wife who had no blot on her escutcheon.* But it is not so. The care in choosing a Regular wife appears to diminish as one rises in the social scale. Nothing would induce an aspiring Isosceles, who had hopes of generating an Equilateral Son, to take a wife who reckoned a single Irregularity among her Ancestors; a Square or Pentagon, who is confident that his family is steadily on the rise, does not enquire above the five-hundredth generation; a Hexagon or Dodecagon is even more careless of the wife's pedigree; but a Circle has been known deliberately to take a wife who has had an Irregular Great-Grandfather, and all because of some slight superiority of lustre, or because of the charms of a low voice—which, with us, even more than with you, is thought 'an excellent thing in Woman.'*

Such ill-judged marriages are, as might be expected, barren, if they do not result in positive Irregularity or in diminution of sides; but none of these evils have hitherto proved sufficiently deterrent. The loss of a few sides in a highly-developed Polygon is not easily noticed, and is sometimes compensated by a successful operation in the Neo-Therapeutic Gymnasium, as I have described above; and the Circles are too much disposed to acquiesce in infecundity as a Law of the superior development. Yet, if this evil be not arrested, the gradual diminution of the Circular class may soon become more rapid, and the time may be not far distant when, the race being no longer able to produce a Chief Circle, the Constitution of Flatland must fall.

One other word of warning suggests itself to me, though I cannot so easily mention a remedy; and this also refers to our relations with Women. About three hundred years ago, it was decreed by the Chief Circle that, since Women are deficient in Reason but abundant in Emotion,* they ought no longer to be treated as rational, nor receive any mental education. The consequence was that they were no longer taught to read, nor even to master Arithmetic enough to enable them to count the angles of their husband or children; and hence they sensibly declined during each generation in intellectual power. And this system of female non-education or quietism still prevails.

My fear is that, with the best intentions, this policy has been carried so far as to react injuriously on the Male Sex.

For the consequence is that, as things now are, we Males have to lead a kind of bi-lingual, and I may almost say bi-mental existence. With the Women, we speak of 'love,' 'duty,' 'right,' 'wrong,' 'pity,' 'hope,' and other irrational and emotional conceptions, which have no existence, and the fiction of which has no object except to control feminine exuberances; but among ourselves, and in our books, we have an entirely different vocabulary and I may almost say, idiom. 'Love' then becomes 'the anticipation of benefits;' 'duty' becomes 'necessity' or 'fitness;' and other words are correspondingly transmuted. Moreover, among Women, we use language implying the utmost deference for their Sex; and they fully believe that the Chief Circle Himself is not more devoutly adored by us than they are: but behind their backs they are both regarded and spoken of—by all except the very young—as being little better than 'mindless organisms.'

Our Theology also in the Women's chambers is entirely different from our Theology elsewhere.

Now my humble fear is that this double training, in language as well as in thought, imposes somewhat too heavy a burden upon the young, especially when, at the age of three years old, they are taken from the maternal care and taught to unlearn the old language—except for the purpose of repeating it in the presence of their Mothers and Nurses—and to learn the vocabulary and idiom of Science. Already methinks I discern a weakness in the grasp of mathematical truth at the present time as compared with the more robust intellect of our ancestors three hundred years ago. I say nothing of the possible danger if a Woman should ever surreptitiously learn to read and convey to her Sex the result of her perusal of a single popular volume; nor of the possibility that the indiscretion or disobedience of some infant Male might reveal to a Mother the secrets of the logical dialect. On the simple ground of the enfeebling of the Male intellect, I rest this humble appeal to the highest Authorities to reconsider the regulations of Female Education.

PART II

OTHER WORLDS

'O brave new worlds,
That have such people in them!' *

13

How I had a Vision of Lineland

IT was the last day but one of the 1999th year of our era, and the first day of the Long Vacation.* Having amused myself till a late hour with my favourite recreation of Geometry, I had retired to rest with an unsolved problem in my mind. In the night I had a dream.

I saw before me a vast multitude of small Straight Lines (which I naturally assumed to be Women) interspersed with other Beings still smaller and of the nature of lustrous Points—all moving to and fro in one and the same Straight Line, and, as nearly as I could judge, with the same velocity.

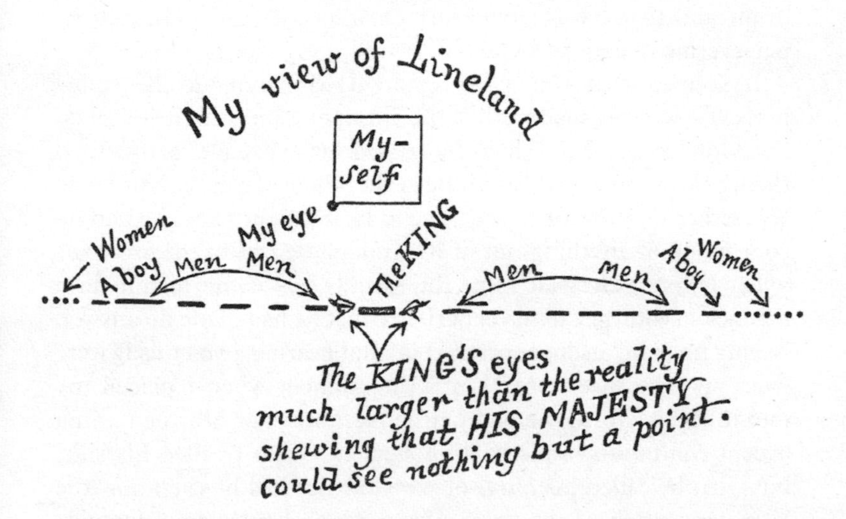

A noise of confused, multitudinous chirping or twittering issued from them at intervals as long as they were moving; but sometimes they ceased from motion, and then all was silence.

Approaching one of the largest of what I thought to be Women, I accosted her, but received no answer. A second and a third appeal on my part were equally ineffectual. Losing patience at what appeared to me intolerable rudeness, I brought my mouth into a position full in front of her mouth so as to intercept her motion, and loudly repeated my question, 'Woman, what signifies this concourse, and this strange and confused chirping, and this monotonous motion to and fro in one and the same Straight Line?'

'I am no Woman,' replied the small Line; 'I am the Monarch of the world. But thou, whence intrudest thou into my realm of Lineland?' Receiving this abrupt reply, I begged pardon if I had in any way startled or molested his Royal Highness; and describing myself as a stranger I besought the King to give me some account of his dominions. But I had the greatest possible difficulty in obtaining any information on points that really interested me; for the Monarch could not refrain from constantly assuming that whatever was familiar to him must also be known to me and that I was simulating ignorance in jest. However, by persevering questions I elicited the following facts:

It seemed that this poor ignorant Monarch—as he called himself—was persuaded that the Straight Line which he called his Kingdom, and in which he passed his existence, constituted the whole of the world, and indeed the whole of Space. Not being able either to move or to see, save in his Straight Line, he had no conception of anything out of it. Though he had heard my voice when I first addressed him, the sounds had come to him in a manner so contrary to his experience that he had made no answer, 'seeing no man,' as he expressed it, 'and hearing a voice as it were from my own intestines.' Until the moment when I placed my mouth in his World, he had neither seen me, nor heard anything except confused sounds beating against—what I called his side, but what he called his *inside* or *stomach;* nor had he even now the least conception of the region from which I had come. Outside his World, or Line, all was a blank to him; nay, not even a blank, for a blank implies Space; say, rather, all was non-existent.

His subjects—of whom the small Lines were Men and the

Points Women—were all alike confined in motion and eye-sight to that single Straight Line, which was their World. It need scarcely be added that the whole of their horizon was limited to a Point; nor could any one ever see anything but a Point. Man, woman, child, thing—each was a Point to the eye of a Linelander. Only by the sound of the voice could sex or age be distinguished. Moreover, as each individual occupied the whole of the narrow path, so to speak, which constituted his Universe, and no one could move to the right or left to make way for passers by, it followed that no Linelander could ever pass another. Once neighbours, always neighbours. Neighbourhood with them was like marriage with us. Neighbours remained neighbours till death did them part.

Such a life, with all vision limited to a Point, and all motion to a Straight Line, seemed to me inexpressibly dreary; and I was surprised to note the vivacity and cheerfulness of the King. Wondering whether it was possible, amid circumstances so unfavourable to domestic relations, to enjoy the pleasures of conjugal union, I hesitated for some time to question his Royal Highness on so delicate a subject; but at last I plunged into it by abruptly inquiring as to the health of his family. 'My wives and children,' he replied, 'are well and happy.'

Staggered at this answer—for in the immediate proximity of the Monarch (as I had noted in my dream before I entered Lineland) there were none but Men—I ventured to reply, 'Pardon me, but I cannot imagine how your Royal Highness can at any time either see or approach their Majesties, when there are at least half a dozen intervening individuals, whom you can neither see through, nor pass by? Is it possible that in Lineland proximity is not necessary for marriage and for the generation of children?'

'How can you ask so absurd a question?' replied the Monarch. 'If it were indeed as you suggest, the Universe would soon be depopulated. No, no; neighbourhood is needless for the union of hearts; and the birth of children is too important a matter to have been allowed to depend upon such an accident as proximity. You cannot be ignorant of this. Yet since you are pleased to affect ignorance, I will instruct you as if you were the veriest baby in

Lineland. Know, then, that marriages are consummated by means of the faculty of sound and the sense of hearing.

'You are of course aware that every Man has two mouths or voices—as well as two eyes—a bass at one, and a tenor at the other, of his extremities. I should not mention this, but that I have been unable to distinguish your tenor in the course of our conversation.' I replied that I had but one voice, and that I had not been aware that His Royal Highness had two. 'That confirms my impression,' said the King, 'that you are not a Man, but a feminine Monstrosity with a bass voice and an utterly uneducated ear. But to continue.

'Nature herself having ordained that every Man should wed two wives—' 'Why two?' asked I. 'You carry your affected simplicity too far,' he cried. 'How can there be a completely harmonious union without the combination of the Four in One, viz. the Bass and Tenor of the Man and the Soprano and Contralto of the two Women?' 'But supposing,' said I, 'that a man should prefer one wife, or three?' 'It is impossible,' he said; 'it is as inconceivable as that two and one should make five, or that the human eye should see a Straight Line.' I would have interrupted him; but he proceeded as follows:

'Once in the middle of each week a Law of Nature compels us to move to and fro with a rhythmic motion of more than usual violence, which continues for the time you would take to count a hundred and one. In the midst of this choral dance, at the fifty-first pulsation, the inhabitants of the Universe pause in full career, and each individual sends forth his richest, fullest, sweetest strain. It is in this decisive moment that all our marriages are made. So exquisite is the adaptation of Bass to Treble, of Tenor to Contralto, that often times the Loved Ones, though twenty thousand leagues away, recognise at once the responsive note of their destined Lover; and, penetrating the paltry obstacles of distance, Love unites the three. The marriage in that instant consummated results in a threefold Male and Female offspring which takes its place in Lineland.'

'What! Always threefold?' said I. 'Must one wife then always have twins?'

'Bass-voiced Monstrosity! yes,' replied the King. 'How else could the balance of the Sexes be maintained, if two girls were not born for every boy? Would you ignore the very Alphabet of Nature?' He ceased, speechless for fury; and some time elapsed before I could induce him to resume his narrative.

'You will not, of course, suppose that every bachelor among us finds his mates at the first wooing in this universal Marriage Chorus. On the contrary, the process is by most of us many times repeated. Few are the hearts whose happy lot it is at once to recognise in each other's voices the partner intended for them by Providence, and to fly into a reciprocal and perfectly harmonious embrace. With most of us the courtship is of long duration. The Wooer's voices may perhaps accord with one of the future wives, but not with both; or not, at first, with either; or the Soprano and Contralto may not quite harmonise. In such cases Nature has provided that every weekly Chorus shall bring the three Lovers into closer harmony. Each trial of voice, each fresh discovery of discord, almost imperceptibly induces the less perfect to modify his or her vocal utterance so as to approximate to the more perfect. And after many trials and many approximations, the result is at last achieved. There comes a day at last, when, while the wonted Marriage Chorus goes forth from universal Lineland, the three far-off Lovers suddenly find themselves in exact harmony, and, before they are aware, the wedded Triplet is rapt vocally into a duplicate embrace; and Nature rejoices over one more marriage and over three more births.'

14

How I vainly tried to explain the nature of Flatland

THINKING that it was time to bring down the Monarch from his raptures to the level of common sense, I determined to endeavour to open up to him some glimpses of the truth, that is to say of the nature of things in Flatland. So I began thus: 'How does your Royal Highness distinguish the shapes and positions of his subjects? I for my part noticed by the sense of sight, before I entered your Kingdom, that some of your people are Lines and others Points, and that some of the Lines are larger—' 'You speak of an impossibility,' interrupted the King; 'you must have seen a vision; for to detect the difference between a Line and a Point by the sense of sight is, as every one knows, in the nature of things, impossible; but it can be detected by the sense of hearing, and by the same means my shape can be exactly ascertained. Behold me—I am a Line, the longest in Lineland, over six inches of Space—' 'Of Length,' I ventured to suggest. 'Fool,' said he, 'Space is Length. Interrupt me again, and I have done.'

I apologised; but he continued scornfully, 'Since you are impervious to argument, you shall hear with your ears how by means of my two voices I reveal my shape to my Wives, who are at this moment six thousand miles seventy yards two feet eight inches away the one to the North, the other to the South. Listen, I call to them.'

He chirruped, and then complacently continued: 'My Wives, at this moment receiving the sound of one of my voices, closely followed by the other, and perceiving that the latter reaches them after an interval in which sound can traverse 6.457 inches, infer that one of my mouths is 6.457 inches further from them than the other, and accordingly know my shape to be 6.457 inches. But you will of course understand that my Wives do not make this calculation every time they hear my two voices. They made it, once for all, before we were married. But they *could* make it at any time. And in the same way I can estimate the shape of any of my Male subjects by the sense of sound.'

'But how,' said I, 'if a Man feigns a Woman's voice with one of his two voices, or so disguises his Southern voice that it cannot be recognised as the echo of the Northern? May not such deceptions cause great inconvenience? And have you no means of checking frauds of this kind by commanding your neighbouring subjects to feel one another?' This of course was a very stupid question; for feeling could not have answered the purpose: but I asked with the view of irritating the Monarch, and I succeeded perfectly.

'What!' cried he in horror, 'explain your meaning.' 'Feel, touch, come into contact,' I replied. 'If you mean by *feeling*,' said the King, 'approaching so close as to leave no space between two individuals, know, Stranger, that this offence is punishable in my dominions by death. And the reason is obvious. The frail form of a Woman, being liable to be shattered by such an approximation, must be preserved by the State; but since Women cannot be distinguished by the sense of sight from Man, the Law ordains universally that neither Man nor Woman shall be approached so closely as to destroy the interval between the approximator and the approximated.

'And indeed what possible purpose would be served by this illegal and unnatural excess of approximation which you call *touching*, when all the ends of so brutal and coarse a process are attained at once more easily and more exactly by the sense of hearing. As to your suggested danger of deception, it is non-existent: for the Voice, being the essence of one's Being, cannot be thus changed at will. But come, suppose that I had the power of passing through solid things, so that I could penetrate my subjects, one after another, even to the number of a billion, verifying the size and distance of each by the sense of *feeling:* how much time and energy would be wasted in this clumsy and inaccurate method! Whereas now, in one moment of audition, I take as it were the census and statistics, local, corporal, mental, and spiritual, of every living being in Lineland. Hark, only hark!'

So saying he paused and listened, as if in an ecstasy, to a sound which seemed to me no better than a tiny chirping from an innumerable multitude of lilliputian grasshoppers.*

'Truly,' replied I, 'your sense of hearing serves you in good stead, and fills up many of your deficiencies. But permit me to point out that your life in Lineland must be deplorably dull. To see nothing but a Point! Not even to be able to contemplate a Straight Line! Nay, not even to know what a Straight Line is! To see, yet to be cut off from those Linear prospects which are vouchsafed to us in Flatland! Better surely to have no sense of sight at all than to see so little! I grant you I have not your discriminative faculty of hearing; for the concert of all Lineland which gives you such intense pleasure, is to me no better than a multitudinous twittering or chirping. But at least I can discern, by sight, a Line from a Point. And let me prove it. Just before I came into your kingdom, I saw you dancing from left to right, and then from right to left, with seven Men and a Woman in your immediate proximity on the left, and eight Men and two Women on your right. Is not this correct?'

'It is correct,' said the King, 'so far as the numbers and sexes are concerned, though I know not what you mean by "right" and "left." But I deny that you saw these things. For how could you see the Line, that is to say the inside, of any Man? But you must have heard these things, and then dreamed that you saw them. And let me ask what you mean by those words "left" and "right." I suppose it is your way of saying Northward and Southward.'

'Not so,' replied I; 'besides your motion of Northward and Southward, there is another motion which I call from right to left.'

King. Exhibit to me, if you please, this motion from left to right.

I. Nay, that I cannot do, unless you could step out of your Line altogether.

King. Out of my Line? Do you mean out of the World? Out of Space?

I. Well, yes. Out of *your* World. Out of *your* Space. For your

Space is not the true Space. True Space is a Plane; but your Space is only a Line.

King. If you cannot indicate this motion from left to right by yourself moving in it, then I beg you to describe it to me in words.

I. If you cannot tell your right side from my left, I fear that no words of mine can make my meaning clear to you. But surely you cannot be ignorant of so simple a distinction.

King. I do not in the least understand you.

I. Alas! How shall I make it clear? When you move straight on, does it not sometimes occur to you that you *could* move in some other way, turning your eye round so as to look in the direction towards which your side is now fronting? In other words, instead of always moving in the direction of one of your extremities, do you never feel a desire to move in the direction, so to speak, of your side?

King. Never. And what do you mean? How can a man's inside 'front' in any direction? Or how can a man move in the direction of his inside?

I. Well then, since words cannot explain the matter, I will try deeds, and will move gradually out of Lineland in the direction which I desire to indicate to you.

At the word I began to move my body out of Lineland. As long as any part of me remained in his dominion and in his view, the King kept exclaiming, 'I see you, I see you still; you are not moving.' But when I had at last moved myself out of his Line, he cried in his shrillest voice, 'She is vanished; she is dead.' 'I am not dead,' replied I; 'I am simply out of Lineland, that is to say, out of the Straight Line which you call Space, and in the true Space, where I can see things as they are. And at this moment I can see

your Line, or side—or inside as you are pleased to call it; and I can also see the Men and Women on the North and South of you, whom I will now enumerate, describing their order, their size, and the interval between each.'

When I had done this at great length, I cried triumphantly, 'Does this at last convince you?' And, with that, I once more entered Lineland, taking up the same position as before.

But the Monarch replied, 'If you were a Man of sense—though, as you appear to have only one voice I have little doubt you are not a Man but a Woman—but, if you had a particle of sense, you would listen to reason. You ask me to believe that there is another Line besides that which my senses indicate, and another motion besides that of which I am daily conscious. I, in return, ask you to describe in words or indicate by motion that other Line of which you speak. Instead of moving, you merely exercise some magic art of vanishing and returning to sight; and instead of any lucid description of your new World, you simply tell me the numbers and sizes of some forty of my retinue, facts known to any child in my capital. Can anything be more irrational or audacious? Acknowledge your folly or depart from my dominions.'

Furious at his perversity, and especially indignant that he professed to be ignorant of my Sex, I retorted in no measured terms, 'Besotted Being! You think yourself the perfection of existence, while you are in reality the most imperfect and imbecile. You profess to see, whereas you can see nothing but a Point! You plume yourself on inferring the existence of a Straight Line; but I *can see* Straight Lines and infer the existence of Angles, Triangles, Squares, Pentagons, Hexagons, and even Circles. Why waste more words? Suffice it that I am the completion of your incomplete self. You are a Line, but I am a Line of Lines, called in my country a Square: and even I, infinitely superior though I am to you, am of little account among the great Nobles of Flatland, whence I have come to visit you, in the hope of enlightening your ignorance.'

Hearing these words the King advanced towards me with a menacing cry as if to pierce me through the diagonal; and in that same moment there arose from myriads of his subjects a multitudinous war-cry, increasing in vehemence till at last methought it rivalled the roar of an army of a hundred thousand Isosceles, and the artillery of a thousand Pentagons. Spell-bound and motionless, I could neither speak nor move to avert the impending destruction; and still the noise grew louder, and the King came closer, when I awoke to find the breakfast-bell recalling me to the realities of Flatland.

15

Concerning a Stranger from Spaceland

FROM dreams I proceed to facts.

It was the last day of the 1999th year of our era. The pattering of the rain had long ago announced nightfall; and I was sitting[1] in the company of my wife, musing on the events of the past and the prospects of the coming year, the coming century, the coming Millennium.

My four Sons and two orphan Grandchildren had retired to their several apartments; and my Wife alone remained with me to see the old Millennium out and the new one in.

I was rapt in thought, pondering in my mind some words that had casually issued from the mouth of my youngest Grandson, a most promising young Hexagon of unusual brilliancy and perfect angularity. His uncles and I had been giving him his usual practical lesson in Sight Recognition, turning ourselves upon our centres, now rapidly, now more slowly, and questioning him as to our positions; and his answers had been so satisfactory that I had been induced to reward him by giving him a few hints on Arithmetic as applied to Geometry.

Taking nine Squares, each an inch every way, I had put them together so as to make one large Square, with a side of three inches, and I had hence proved to my little Grandson that—though it was impossible for us to *see* the inside of the Square—yet we might ascertain the number of square inches in a Square by simply squaring the number of inches in the side: 'and thus,'

[1] When I say 'sitting,' of course I do not mean any change of attitude such as you in Spaceland signify by that word; for as we have no feet, we can no more 'sit' nor 'stand' (in your sense of the word) than one of your soles or flounders.

Nevertheless, we perfectly well recognise the different mental states of volition implied in 'lying,' 'sitting,' and 'standing,' which are to some extent indicated to a beholder by a slight increase of lustre corresponding to the increase of volition.

But on this, and a thousand other kindred subjects, time forbids me to dwell.

said I, 'we know that 3^2, or 9, represents the number of square inches in a Square whose side is 3 inches long.'

The little Hexagon meditated on this a while and then said to me: 'But you have been teaching me to raise numbers to the third power; I suppose 3^3 must mean something in Geometry; what does it mean?' 'Nothing at all,' replied I, 'not at least in Geometry; for Geometry has only Two Dimensions.' And then I began to show the boy how a Point by moving through a length of three inches makes a Line of three inches, which may be represented by 3; and how a Line of three inches, moving parallel to itself through a length of three inches, makes a Square of three inches every way, which may be represented by 3^2.

Upon this, my Grandson, again returning to his former suggestion, took me up rather suddenly and exclaimed, 'Well, then, if a Point by moving three inches, makes a Line of three inches represented by 3; and if a straight Line of three inches, moving parallel to itself, makes a Square of three inches every way, represented by 3^2; it must be that a Square of three inches every way, moving somehow parallel to itself (but I don't see how) must make a Something else (but I don't see what) of three inches every way—and this must be represented by 3^3.'

'Go to bed,' said I, a little ruffled by his interruption; 'if you would talk less nonsense, you would remember more sense.'

So my Grandson had disappeared in disgrace; and there I sat by my Wife's side, endeavouring to form a retrospect of the year 1999 and of the possibilities of the year 2000, but not quite able to shake off the thoughts suggested by the prattle of my bright little Hexagon. Only a few sands now remained in the half-hour glass. Rousing myself from my reverie I turned the glass Northward for the last time in the old Millennium; and in the act, I exclaimed aloud, 'The boy is a fool.'

Straightway I became conscious of a Presence in the room, and a chilling breath thrilled through my very being. 'He is no such thing,' cried my Wife, 'and you are breaking the Commandments in thus dishonouring your own Grandson.' But I took no notice of her. Looking round in every direction I could see nothing; yet still I *felt* a Presence, and shivered as the cold whisper came again.

I started up. 'What is the matter?' said my Wife, 'there is no draught; what are you looking for? There is nothing.' There was nothing; and I resumed my seat, again exclaiming, 'The boy is a fool, I say; 3^3 can have no meaning in Geometry.' At once there came a distinctly audible reply, 'The boy is not a fool; and 3^3 has an obvious Geometrical meaning.'

My Wife as well as myself heard the words, although she did not understand their meaning, and both of us sprang forward in the direction of the sound. What was our horror when we saw before us a Figure! At the first glance it appeared to be a Woman, seen sideways; but a moment's observation shewed me that the extremities passed into dimness too rapidly to represent one of the Female Sex; and I should have thought it a Circle, only that it seemed to change its size in a manner impossible for a Circle or for any Regular Figure of which I had had experience.

But my Wife had not my experience, nor the coolness necessary to note these characteristics. With the usual hastiness and unreasoning jealousy of her Sex, she flew at once to the conclusion that a Woman had entered the house through some small aperture. 'How comes this person here?' she exclaimed, 'you promised me, my dear, that there should be no ventilators in our new house.' 'Nor are there any,' said I; 'but what makes you think that the stranger is a Woman? I see by my power of Sight Recognition—' 'Oh, I have no patience with your Sight Recognition,' replied she, ' "Feeling is believing" and "A Straight Line to the touch is worth a Circle to the sight" '—two Proverbs, very common with the Frailer Sex in Flatland.

'Well,' said I, for I was afraid of irritating her, 'if it must be so, demand an introduction.' Assuming her most gracious manner, my Wife advanced towards the Stranger, 'Permit me, Madam, to feel and be felt by—' then, suddenly recoiling, 'Oh! it is not a Woman, and there are no angles either, not a trace of one. Can it be that I have so misbehaved to a perfect Circle?'

'I am indeed, in a certain sense a Circle,' replied the Voice, 'and a more perfect Circle than any in Flatland; but to speak more accurately, I am many Circles in one.' Then he added more mildly, 'I have a message, dear Madam, to your husband, which I must not deliver in your presence; and, if you would suffer us to retire for a few minutes—' But my Wife would not listen to the proposal that our august Visitor should so incommode himself, and assuring the Circle that the hour for her own retirement had long passed, with many reiterated apologies for her recent indiscretion, she at last retreated to her apartment.

I glanced at the half-hour glass. The last sands had fallen. The second Millennium* had begun.

16

How the Stranger vainly endeavoured to reveal to me in words the mysteries of Spaceland

As soon as the sound of the Peace-cry of my departing Wife had died away, I began to approach the Stranger with the intention of taking a nearer view and of bidding him be seated: but his appearance struck me dumb and motionless with astonishment. Without the slightest symptoms of angularity he nevertheless varied every instant with gradations of size and brightness scarcely possible for any Figure within the scope of my experience. The thought flashed across me that I might have before me a burglar or cut-throat, some monstrous Irregular Isosceles, who, by feigning the voice of a Circle, had obtained admission somehow into the house, and was now preparing to stab me with his acute angle.

In a sitting-room, the absence of Fog (and the season happened to be remarkably dry), made it difficult for me to trust to Sight Recognition, especially at the short distance at which I was standing. Desperate with fear, I rushed forward with an unceremonious 'You must permit me, Sir—' and felt him. My Wife was right. There was not the trace of an angle, not the slightest roughness or inequality: never in my life had I met with a more perfect Circle. He remained motionless while I walked round him, beginning from his eye and returning to it again. Circular he was throughout, a perfectly satisfactory Circle; there could not be a doubt of it. Then followed a dialogue, which I will endeavour to set down as near as I can recollect it, omitting only some of my profuse apologies—for I was covered with shame and humiliation that I, a Square, should have been guilty of the impertinence of feeling a Circle. It was commenced by the Stranger with some impatience at the lengthiness of my introductory process.

Stranger. Have you felt me enough by this time? Are you not introduced to me yet?

I. Most illustrious Sir, excuse my awkwardness, which arises not from ignorance of the usages of polite society, but from a little surprise and nervousness, consequent on this somewhat unexpected visit. And I beseech you to reveal my indiscretion to no one, and especially not to my Wife. But before your Lordship enters into further communications, would he deign to satisfy the curiosity of one who would gladly know whence his Visitor came?

Stranger. From Space, from Space, Sir: whence else?

I. Pardon me, my Lord, but is not your Lordship already in Space, your Lordship and his humble servant, even at this moment?

Stranger. Pooh! what do you know of Space? Define Space.

I. Space, my Lord, is height and breadth indefinitely prolonged.

Stranger. Exactly: you see you do not even know what Space is. You think it is of Two Dimensions only; but I have come to announce to you a Third—height, breadth, and length.

I. Your Lordship is pleased to be merry. We also speak of length and height, or breadth and thickness, thus denoting Two Dimensions by four names.

Stranger. But I mean not only three names, but Three Dimensions.

I. Would your Lordship indicate or explain to me in what direction is the Third Dimension, unknown to me?

Stranger. I came from it. It is up above and down below.

I. My Lord means seemingly that it is Northward and Southward.

Stranger. I mean nothing of the kind. I mean a direction in which you cannot look, because you have no eye in your side.

I. Pardon me, my Lord, a moment's inspection will convince your Lordship that I have a perfect luminary at the juncture of two of my sides.

Stranger. Yes: but in order to see into Space you ought to have an eye, not on your Perimeter, but on your side, that is, on what you would probably call your inside; but we in Spaceland should call it your side.

I. An eye in my inside! An eye in my stomach! Your Lordship jests.

Stranger. I am in no jesting humour. I tell you that I come from Space, or, since you will not understand what Space means, from the Land of Three Dimensions whence I but lately looked down upon your Plane which you call Space forsooth. From that position of advantage I discerned all that you speak of as *solid* (by which you mean 'enclosed on four sides'), your houses, your churches, your very chests and safes, yes even your insides and stomachs, all lying open and exposed to my view.

I. Such assertions are easily made, my Lord.

Stranger. But not easily proved, you mean. But I mean to prove mine.

When I descended here, I saw your four Sons, the Pentagons, each in his apartment, and your two Grandsons the Hexagons; I saw your youngest Hexagon remain a while with you and then retire to his room, leaving you and your Wife alone. I saw your Isosceles servants, three in number, in the kitchen at supper, and the little Page in the scullery. Then I came here, and how do you think I came?

I. Through the roof, I suppose.

Stranger. Not so. Your roof, as you know very well, has been recently repaired, and has no aperture by which even a Woman could penetrate. I tell you I come from Space. Are you not convinced by what I have told you of your children and household.

I. Your Lordship must be aware that such facts touching the belongings of his humble servant might be easily ascertained by any one in the neighbourhood possessing your Lordship's ample means of obtaining information.

Stranger. (*To himself*). What must I do?* Stay; one more argument suggests itself to me. When you see a Straight Line—your wife, for example—how many Dimensions do you attribute to her?

I. Your Lordship would treat me as if I were one of the vulgar who, being ignorant of Mathematics, suppose that a Woman is really a Straight Line, and only of One Dimension. No, no, my Lord; we Squares are better advised, and are as well aware as your

Lordship that a Woman, though popularly called a Straight Line, is, really and scientifically, a very thin Parallelogram, possessing Two Dimensions, like the rest of us, viz., length and breadth (or thickness).

Stranger. But the very fact that a Line is visible implies that it possesses yet another Dimension.

I. My Lord, I have just acknowledged that a Woman is broad as well as long. We see her length, we infer her breadth; which, though very slight, is capable of measurement.

Stranger. You do not understand me. I mean that when you see a Woman, you ought—besides inferring her breadth—to see her length, and to *see* what we call her *height*; although that last Dimension is infinitesimal in your country. If a line were mere length without 'height,' it would cease to occupy space and would become invisible. Surely you must recognize this?

I. I must indeed confess that I do not in the least understand your Lordship. When we in Flatland see a Line, we see length and *brightness*. If the brightness disappears, the line is extinguished, and, as you say, ceases to occupy space. But am I to suppose that your Lordship gives to brightness the title of a Dimension, and that what we call 'bright' you call 'high'?

Stranger. No, indeed. By 'height' I mean a Dimension like your length: only, with you, 'height' is not so easily perceptible, being extremely small.

I. My Lord, your assertion is easily put to the test. You say I have a Third Dimension, which you call 'height.' Now, Dimension implies direction and measurement. Do but measure my 'height,' or merely indicate to me the direction in which my 'height' extends, and I will become your convert. Otherwise, your Lordship's own understanding must hold me excused.

Stranger. (*To himself*). I can do neither. How shall I convince him? Surely a plain statement of facts followed by ocular demonstration ought to suffice.—Now, Sir; listen to me.

You are living on a Plane. What you style Flatland is the vast level surface of what I may call a fluid, on, or in, the top of which you and your countrymen move about, without rising above it or falling below it.

I am not a plane Figure, but a Solid. You call me a Circle; but in reality I am not a Circle, but an infinite number of Circles, of size varying from a Point to a Circle of thirteen inches in diameter, one placed on the top of the other. When I cut through your plane as I am now doing, I make in your plane a section which you, very rightly, call a Circle. For even a Sphere—which is my proper name in my own country—if he manifest himself at all to an inhabitant of Flatland—must needs manifest himself as a Circle.

Do you not remember—for I, who see all things, discerned last night the phantasmal vision of Lineland written upon your brain—do you not remember, I say, how, when you entered the realm of Lineland, you were compelled to manifest yourself to the King not as a Square, but as a Line, because that Linear Realm had not Dimensions enough to represent the whole of you, but only a slice or section of you? In precisely the same way, your country of Two Dimensions is not spacious enough to represent me, a being of Three, but can only exhibit a slice or section of me, which is what you call a Circle.

The diminished brightness of your eye indicates incredulity. But now prepare to receive proof positive of the truth of my assertions. You cannot indeed see more than one of my sections, or Circles, at a time; for you have no power to raise your eye out of the plane of Flatland; but you can at least see that, as I rise in Space, so my section becomes smaller. See now, I will rise; and the effect upon your eye will be that my Circle will become smaller and smaller till it dwindles to a point and finally vanishes.

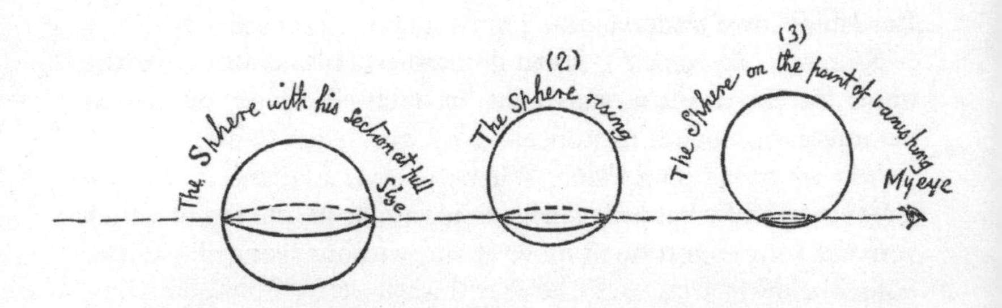

There was no 'rising' that I could see; but he diminished and finally vanished. I winked once or twice to make sure that I was not dreaming. But it was no dream. For from the depths of nowhere came forth a hollow voice—close to my heart it seemed—'Am I quite gone? Are you convinced now? Well, now I will gradually return to Flatland, and you shall see my section become larger and larger.'

Every reader in Spaceland will easily understand that my mysterious Guest was speaking the language of truth and even of simplicity. But to me, proficient though I was in Flatland Mathematics, it was by no means a simple matter. The rough diagram given above will make it clear to any Spaceland child that the Sphere, ascending in the three positions indicated there, must needs have manifested himself to me, or to any Flatlander, as a Circle, at first of full size, then small, and at last very small indeed, approaching to a Point. But to me, although I saw the facts before me, the causes were as dark as ever. All that I could comprehend was, that the Circle had made himself smaller and vanished, and that he had now reappeared and was rapidly making himself larger.

When he had regained his original size, he heaved a deep sigh; for he perceived by my silence that I had altogether failed to comprehend him. And indeed I was now inclining to the belief that he must be no Circle at all, but some extremely clever juggler;* or else that the old wives' tales were true, and that after all there were such people as Enchanters and Magicians.

After a long pause he muttered to himself, 'One resource alone remains, if I am not to resort to action. I must try the method of Analogy.' Then followed a still longer silence, after which he continued our dialogue.

Sphere. Tell me, Mr. Mathematician; if a Point moves Northward, and leaves a luminous wake, what name would you give to the wake?

I. A straight Line.

Sphere. And a straight Line has how many extremities?

I. Two.

Sphere. Now conceive the Northward straight line moving

parallel to itself, East and West, so that every point in it leaves behind it the wake of a straight Line. What name will you give to the Figure thereby formed? We will suppose that it moves through a distance equal to the original straight Line.—What name, I say?

I. A Square.

Sphere. And how many sides has a Square? And how many angles?

I. Four sides and four angles.

Sphere. Now stretch your imagination a little, and conceive a Square in Flatland, moving parallel to itself upward.

I. What? Northward?

Sphere. No, not Northward; upward; out of Flatland altogether.

If it moved Northward, the Southern points in the Square would have to move through the positions previously occupied by the Northern points. But that is not my meaning.

I mean that every Point in you—for you are a Square and will serve the purpose of my illustration—every Point in you, that is to say in what you call your inside, is to pass upwards through Space in such a way that no Point shall pass through the position previously occupied by any other Point; but each Point shall describe a straight Line of its own. This is all in accordance with Analogy; surely it must be clear to you.

Restraining my impatience—for I was now under a strong temptation to rush blindly at my Visitor and to precipitate him into Space, or out of Flatland, anywhere, so that I could get rid of him—I replied:—

'And what may be the nature of the Figure which I am to shape out by this motion which you are pleased to denote by the word "upward"? I presume it is describable in the language of Flatland.'

Sphere. Oh, certainly. It is all plain and simple, and in strict accordance with Analogy—only, by the way, you must not speak of the result as being a Figure, but as a Solid. But I will describe it to you. Or rather not I, but Analogy.

We began with a single Point, which of course—being itself a Point—has only *one* terminal Point.

One Point produces a Line with *two* terminal Points.

One Line produces a Square with *four* terminal Points.

Now you can yourself give the answer to your own question: 1, 2, 4, are evidently in Geometrical Progression. What is the next number.

I. Eight.

Sphere. Exactly. The one Square produces a *Something-which-you-do-not-as-yet-know-a-name-for-but-which-we-call-a-Cube* with *eight* terminal Points. Now are you convinced?

I. And has this Creature sides, as well as angles or what you call 'terminal Points?'

Sphere. Of course; and all according to Analogy. But, by the way, not what *you* call sides, but what *we* call sides. You would call them *solids*.

I. And how many solids or sides will appertain to this Being whom I am to generate by the motion of my inside in an 'upward' direction, and whom you call a Cube?

Sphere. How can you ask? And you a mathematician! The side of anything is always, if I may so say, one Dimension behind the thing. Consequently, as there is no Dimension behind a Point, a Point has 0 sides; a Line, if I may so say, has 2 sides (for the Points of a Line may be called by courtesy, its sides); a Square has 4 sides; 0, 2, 4; what Progression do you call that?

I. Arithmetical.

Sphere. And what is the next number?

I. Six.

Sphere. Exactly. Then you see you have answered your own question. The Cube which you will generate will be bounded by six sides, that is to say, six of your insides. You see it all now, eh?

'Monster,' I shrieked, 'be thou juggler, enchanter, dream, or devil, no more will I endure thy mockeries. Either thou or I must perish.' And saying these words I precipitated myself upon him.

17

How the Sphere, having in vain tried words, resorted to deeds

It was in vain. I brought my hardest right angle into violent collision with the Stranger, pressing on him with a force sufficient to have destroyed any ordinary Circle: but I could feel him slowly and unarrestably slipping from my contact; not edging to the right nor to the left, but moving somehow out of the world and vanishing to nothing. Soon there was a blank. But I still heard the Intruder's voice.

Sphere. Why will you refuse to listen to reason? I had hoped to find in you—as being a man of sense and an accomplished mathematician—a fit apostle for the Gospel of the Three Dimensions, which I am allowed to preach once only in a thousand years: but now I know not how to convince you. Stay, I have it. Deeds, and not words, shall proclaim the truth. Listen, my friend.

I have told you I can see from my position in Space the inside of all things that you consider closed. For example, I see in yonder cupboard near which you are standing, several of what you call boxes (but like everything else in Flatland, they have no tops nor bottoms) full of money; I see also two tablets of accounts. I am about to descend into that cupboard and to bring you one of those tablets. I saw you lock the cupboard half an hour ago, and I know you have the key in your possession. But I descend from Space; the doors, you see, remain unmoved. Now I am in the cupboard and am taking the tablet. Now I have it. Now I ascend with it.

I rushed to the closet and dashed the door open. One of the tablets was gone. With a mocking laugh, the Stranger appeared in the other corner of the room, and at the same time the tablet appeared upon the floor. I took it up. There could be no doubt—it was the missing tablet.

I groaned with horror, doubting whether I was not out of my senses; but the Stranger continued: 'Surely you must now see that my explanation, and no other, suits the phenomena. What you call Solid things are really superficial; what you call Space is really nothing but a great Plane. I am in Space, and look down upon the insides of the things of which you only see the outsides. You could leave this Plane yourself, if you could but summon up the necessary volition. A slight upward or downward motion would enable you to see all that I can see.

'The higher I mount, and the further I go from your Plane, the more I can see, though of course I see it on a smaller scale. For example, I am ascending; now I can see your neighbour the Hexagon and his family in their several apartments; now I see the inside of the Theatre, ten doors off, from which the audience is only just departing; and on the other side a Circle in his study, sitting at his books. Now I shall come back to you. And, as a crowning proof, what do you say to my giving you a touch, just the least touch, in your stomach? It will not seriously injure you, and the slight pain you may suffer cannot be compared with the mental benefit you will receive.'

Before I could utter a word of remonstrance, I felt a shooting pain in my inside, and a demoniacal laugh seemed to issue from within me. A moment afterwards the sharp agony had ceased, leaving nothing but a dull ache behind, and the Stranger began to reappear, saying, as he gradually increased in size, 'There, I have not hurt you much, have I? If you are not convinced now, I don't know what will convince you. What say you?'

My resolution was taken. It seemed intolerable that I should endure existence subject to the arbitrary visitations of a Magician who could thus play tricks with one's very stomach. If only I could in any way manage to pin him against the wall till help came!

Once more I dashed my hardest angle against him, at the same time alarming the whole household by my cries for aid. I believe, at the moment of my onset, the Stranger had sunk below our Plane, and really found difficulty in rising. In any case he remained motionless, while I, hearing, as I thought, the sound of some help approaching, pressed against him with redoubled vigour, and continued to shout for assistance.

A convulsive shudder ran through the Sphere. 'This must not be,' I thought I heard him say; 'either he must listen to reason, or I must have recourse to the last resource of civilization.' Then, addressing me in a louder tone, he hurriedly exclaimed, 'Listen: no stranger must witness what you have witnessed. Send your Wife back at once, before she enters the apartment. The Gospel of Three Dimensions must not be thus frustrated. Not thus must the fruits of one thousand years of waiting be thrown away. I hear her coming. Back! back! Away from me, or you must go with me—whither you know not—into the Land of Three Dimensions!'

'Fool! Madman! Irregular!' I exclaimed; 'never will I release thee; thou shalt pay the penalty of thine impostures.'

'Ha! Is it come to this?' thundered the Stranger: 'then meet your fate: out of your Plane you go. Once, twice, thrice! 'Tis done!'

18

How I came to Spaceland, and what I saw there

An unspeakable horror seized me. There was a darkness; then a dizzy, sickening sensation of sight that was not like seeing; I saw a Line that was no Line; Space that was not Space; I was myself, and not myself. When I could find voice, I shrieked aloud in agony, 'Either this is madness or it is Hell.' 'It is neither,' calmly replied the voice of the Sphere, 'it is Knowledge; it is Three Dimensions: open your eye once again and try to look steadily.'

I looked, and, behold, a new world! There stood before me, visibly incorporate, all that I had before inferred, conjectured, dreamed, of perfect Circular beauty. What seemed the centre of the Stranger's form lay open to my view: yet I could see no heart, nor lungs, nor arteries, only a beautiful harmonious Something—for which I had no words; but you, my Readers in Spaceland, would call it the surface of the Sphere.

Prostrating myself mentally before my Guide, I cried, 'How is it, O divine ideal of consummate loveliness and wisdom, that I see thy inside, and yet cannot discern thy heart, thy lungs, thy arteries, thy liver?' 'What you think you see, you see not,' he replied; 'it is not given to you, nor to any other Being, to behold my internal parts. I am of a different order of Beings from those in Flatland. Were I a Circle, you could discern my intestines, but I am a Being composed, as I told you before, of many Circles, the Many in the One, called in this country a Sphere. And, just as the outside of a Cube is a Square, so the outside of a Sphere presents the appearance of a Circle.'

Bewildered though I was by my Teacher's enigmatic utterance, I no longer chafed against it, but worshipped him in silent adoration. He continued, with more mildness in his voice: 'Distress not yourself if you cannot at first understand the deeper mysteries of Spaceland. By degrees they will dawn upon you. Let us begin by casting back a glance at the region whence you came. Return with me a while to the plains of Flatland, and I will show you that which you have so often reasoned and thought about, but never seen with the sense of sight—a visible angle.' 'Impossible!' I cried; but, the Sphere leading the way, I followed as if in a dream, till once more his voice arrested me: 'Look yonder, and behold your own Pentagonal house and all its inmates.'

I looked below, and saw with my physical eye all that domestic individuality which I had hitherto merely inferred with the understanding. And how poor and shadowy was the inferred conjecture in comparison with the reality which I now beheld! My four Sons calmly asleep in the North-Western rooms, my two orphan Grandsons to the South; the Servants, the Butler, my Daughter, all in their several apartments. Only my affectionate Wife, alarmed by my continued absence, had quitted her room and was roving up and down in the Hall, anxiously awaiting my return. Also the Page, aroused by my cries, had left his room, and under pretext of ascertaining whether I had fallen somewhere in a faint, was prying into the cabinet in my study. All this I could now *see*, not merely infer; and as we came nearer and nearer, I could discern even the contents of my cabinet, and the two chests of gold, and the tablets of which the Sphere had made mention.

Touched by my Wife's distress, I would have sprung downward to reassure her, but I found myself incapable of motion. 'Trouble not yourself about your Wife,' said my Guide; 'she will not be long left in anxiety; meantime, let us take a survey of Flatland.'

Once more I felt myself rising through space. It was even as the Sphere had said. The further we receded from the object we beheld, the larger became the field of vision. My native city, with the interior of every house and every creature therein, lay open to my view in miniature. We mounted higher, and lo, the secrets of the earth, the depths of mines and inmost caverns of the hills, were bared before me.

Awestruck at the sight of the mysteries of the earth, thus unveiled before my unworthy eye, I said to my Companion, 'Behold, I am become as a God. For the wise men in our country

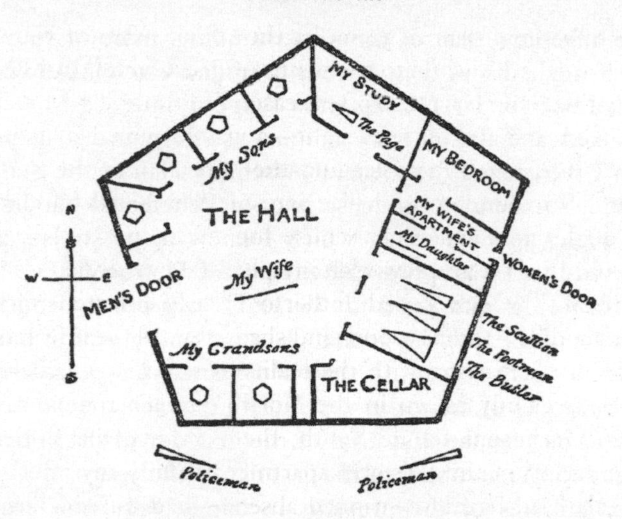

say that to see all things, or as they express it, *omnividence*, is the attribute of God alone.' There was something of scorn in the voice of my Teacher as he made answer: 'Is it so indeed? Then the very pickpockets and cut-throats of my country are to be worshipped by your wise men as being Gods: for there is not one of them that does not see as much as you see now. But trust me, your wise men are wrong.'

I. Then is omnividence the attribute of others beside Gods?

Sphere. I do not know. But, if a pick-pocket or a cut-throat of our country can see everything that is in your country, surely that is no reason why the pick-pocket or cut-throat should be accepted by you as a God. This omnividence, as you call it—it is not a common word in Spaceland—does it make you more just, more merciful, less selfish, more loving? Not in the least. Then how does it make you more divine?

I. 'More merciful, more loving!' But these are the qualities of women! And we know that a Circle is a higher Being than a Straight Line, in so far as knowledge and wisdom are more to be esteemed than mere affection.

Sphere. It is not for me to classify human faculties according to merit. Yet many of the best and wisest in Spaceland think more of the affections than of the understanding, more of your despised Straight Lines than of your belauded Circles. But enough of this. Look yonder. Do you know that building?

I looked, and afar off I saw an immense Polygonal structure, in which I recognized the General Assembly Hall of the States of Flatland, surrounded by dense lines of Pentagonal buildings at right angles to each other, which I knew to be streets; and I perceived that I was approaching the great Metropolis.

'Here we descend,' said my Guide. It was now morning, the first hour of the first day of the two thousandth year of our era. Acting, as was their wont, in strict accordance with precedent, the highest Circles of the realm were meeting in solemn conclave, as they had met on the first hour of the first day of the year 1000, and also on the first hour of the first day of the year 0.

The minutes of the previous meetings were now read by one whom I at once recognized as my brother, a perfectly Symmetrical Square, and the Chief Clerk of the High Council. It was found recorded on each occasion that: 'Whereas the States had been troubled by divers ill-intentioned persons pretending to have received revelations from another World, and professing to produce demonstrations whereby they had instigated to frenzy both themselves and others, it had been for this cause unanimously resolved by the Grand Council that on the first day of each millenary, special injunctions be sent to the Prefects in the several districts of Flatland, to make strict search for such misguided persons, and without formality of mathematical examination, to destroy all such as were Isosceles of any degree, to scourge and imprison any regular Triangle, to cause any Square or Pentagon to be sent to the district Asylum, and to arrest any one of higher rank, sending him straightway to the Capital to be examined and judged by the Council.'

'You hear your fate,' said the Sphere to me, while the Council was passing for the third time the formal resolution. 'Death or imprisonment awaits the Apostle of the Gospel of Three Dimensions.' 'Not so,' replied I, 'the matter is now so clear to me, the nature of real space so palpable, that methinks I could make a child understand it. Permit me but to descend at this moment and enlighten them.' 'Not yet,' said my Guide, 'the time will come for that. Meantime I must perform my mission. Stay thou there in thy place.' Saying these words, he leaped with great dexterity into the sea (if I may so call it) of Flatland, right in the midst of the ring of Counsellors. 'I come,' cried he, 'to proclaim that there is a land of Three Dimensions.'

I could see many of the younger Counsellors start back in manifest horror, as the Sphere's circular section widened before them. But on a sign from the presiding Circle,—who showed not the slightest alarm or surprise—six Isosceles of a low type from six different quarters rushed upon the Sphere. 'We have him,' they cried; 'No; yes; we have him still! he's going! he's gone!'

'My Lords,' said the President to the Junior Circles of the Council, 'there is not the slightest need for surprise; the secret archives, to which I alone have access, tell me that a similar occurrence happened on the last two millennial commencements. You will, of course, say nothing of these trifles outside the Cabinet.'

Raising his voice, he now summoned the guard. 'Arrest the policemen; gag them. You know your duty.' After he had consigned to their fate the wretched policemen—ill-fated and unwilling witnesses of a State-secret which they were not to be permitted to reveal—he again addressed the Counsellors. 'My Lords, the business of the Council being concluded, I have only to wish you a happy New Year.' Before departing, he expressed, at some length, to the Clerk, my excellent but most unfortunate brother, his sincere regret that, in accordance with precedent and for the sake of secrecy, he must condemn him to perpetual imprisonment, but added his satisfaction that, unless some mention were made by him of that day's incident, his life would be spared.

19

How, though the Sphere showed me other mysteries of Spaceland, I still desired more; and what came of it

WHEN I saw my poor brother led away to imprisonment, I attempted to leap down into the Council Chamber, desiring to intercede on his behalf, or at least bid him farewell. But I found that I had no motion of my own. I absolutely depended on the volition of my Guide, who said in gloomy tones, 'Heed not thy brother; haply thou shalt have ample time hereafter to condole with him. Follow me.'

Once more we ascended into space. 'Hitherto,' said the Sphere, 'I have shown you naught save Plane Figures and their interiors. Now I must introduce you to Solids, and reveal to you the plan upon which they are constructed. Behold this multitude of moveable square cards. See, I put one on another, not, as you supposed, Northward of the other, but *on* the other. Now a second, now a third. See, I am building up a Solid by a multitude of Squares parallel to one another. Now the Solid is complete, being as high as it is long and broad, and we call it a Cube.'

'Pardon me, my Lord,' replied I; 'but to my eye the appearance is as of an Irregular Figure whose inside is laid open to the view; in other words, methinks I see no Solid, but a Plane such as we infer in Flatland; only of an Irregularity which betokens some monstrous criminal, so that the very sight of it is painful to my eyes.'

'True,' said the Sphere; 'it appears to you a Plane, because you are not accustomed to light and shade and perspective; just as in Flatland a Hexagon would appear a Straight Line to one who has not the Art of Sight Recognition. But in reality it is a Solid, as you shall learn by the sense of Feeling.'

He then introduced me to the Cube, and I found that this marvellous Being was indeed no Plane, but a Solid; and that he

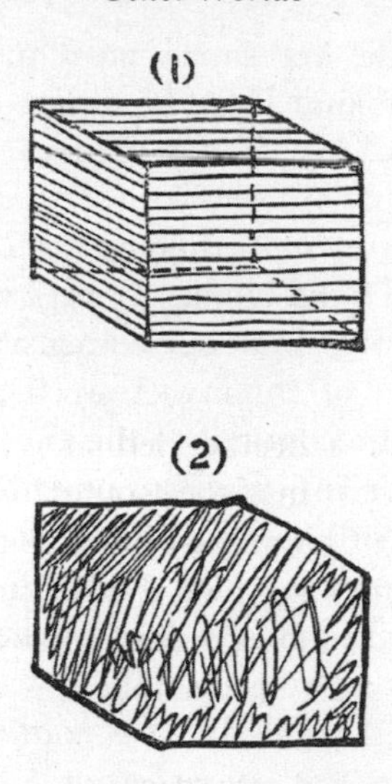

was endowed with six plane sides and eight terminal points called solid angles; and I remembered the saying of the Sphere that just such a Creature as this would be formed by a Square moving, in Space, parallel to himself: and I rejoiced to think that so insignificant a Creature as I could in some sense be called the Progenitor of so illustrious an offspring.

But still I could not fully understand the meaning of what my Teacher had told me concerning 'light' and 'shade' and 'perspective'; and I did not hesitate to put my difficulties before him.

Were I to give the Sphere's explanation of these matters, succinct and clear though it was, it would be tedious to an inhabitant of Space, who knows these things already. Suffice it, that by his lucid statements, and by changing the position of objects and lights, and by allowing me to feel the several objects and even his own sacred Person, he at last made all things clear to me, so that I could now readily distinguish between a Circle and a Sphere, a Plane Figure and a Solid.

This was the Climax, the Paradise, of my strange eventful History. Henceforth I have to relate the story of my miserable

Fall:—most miserable, yet surely most undeserved! For why should the thirst for knowledge be aroused, only to be disappointed and punished! My volition shrinks from the painful task of recalling my humiliation; yet, like a second Prometheus,* I will endure this and worse, if by any means I may arouse in the interiors of Plane and Solid Humanity a spirit of rebellion against the Conceit which would limit our Dimensions to Two or Three or any number short of Infinity. Away then with all personal considerations! Let me continue to the end, as I began, without further digressions or anticipations, pursuing the plain path of dispassionate History. The exact facts, the exact words,—and they are burnt in upon my brain,—shall be set down without alteration of an iota; and let my Readers judge between me and Destiny.

The Sphere would willingly have continued his lessons by indoctrinating me in the conformation of all regular Solids, Cylinders, Cones, Pyramids, Pentahedrons, Hexahedrons, Dodecahedrons and Spheres: but I ventured to interrupt him. Not that I was wearied of knowledge. On the contrary, I thirsted for yet deeper and fuller draughts than he was offering to me.

'Pardon me,' said I, 'O Thou Whom I must no longer address as the Perfection of all Beauty; but let me beg thee to vouchsafe thy servant a sight of thine interior.'

Sphere. 'My what?'

I. 'Thine interior: thy stomach, thy intestines.'

Sphere. 'Whence this ill-timed impertinent request? And what mean you by saying that I am no longer the Perfection of all Beauty?'

I. My Lord, your own wisdom has taught me to aspire to One even more great, more beautiful, and more closely approximate to Perfection than yourself. As you yourself, superior to all Flatland forms, combine many Circles in One, so doubtless there is One above you who combines many Spheres in One Supreme Existence, surpassing even the Solids of Spaceland. And even as we, who are now in Space, look down on Flatland and see the insides of all things, so of a certainty there is yet above us some higher, purer region, whither thou dost surely purpose to lead me—O Thou Whom I shall always call, everywhere and in all Dimensions, my Priest, Philosopher, and Friend—some yet more spacious Spacc, some more dimensionable Dimensionality, from the vantage-ground of which we shall look down together upon the revealed insides of Solid things, and where thine own intestines, and those of thy kindred Spheres, will lie exposed to the view of the poor wandering exile from Flatland, to whom so much has already been vouchsafed.

Sphere. Pooh! Stuff! Enough of this trifling! The time is short, and much remains to be done before you are fit to proclaim the Gospel of Three Dimensions to your blind benighted countrymen in Flatland.

I. Nay, gracious Teacher, deny me not what I know it is in thy power to perform. Grant me but one glimpse of thine interior, and I am satisfied for ever, remaining henceforth thy docile pupil, thy unemancipable slave, ready to receive all thy teachings and to feed upon the words that fall from thy lips.

Sphere. Well, then, to content and silence you, let me say at once, I would show you what you wish if I could; but I cannot. Would you have me turn my stomach inside out to oblige you?

I. But my Lord has shown me the intestines of all my countrymen in the Land of Two Dimensions by taking me with him into the Land of Three. What therefore more easy than now to take his servant on a second journey into the blessed region of the Fourth Dimension, where I shall look down with him once more upon this land of Three Dimensions, and see the inside of every three-dimensioned house, the secrets of the solid earth, the treasures of the mines in Spaceland, and the intestines of every solid living creature, even of the noble and adorable Spheres.

Sphere. But where is this land of Four Dimensions?

I. I know not: but doubtless my Teacher knows.

Sphere. Not I. There is no such land. The very idea of it is utterly inconceivable.

I. Not inconceivable, my Lord, to me,* and therefore still less inconceivable to my Master. Nay, I despair not that, even here, in this region of Three Dimensions, your Lordship's art may make the Fourth Dimension visible to me; just as in the Land of Two Dimensions my Teacher's skill would fain have opened the eyes of his blind servant to the invisible presence of a Third Dimension, though I saw it not.

Let me recall the past. Was I not taught below that when I saw a Line and inferred a Plane, I in reality saw a Third unrecognized Dimension, not the same as brightness, called 'height'? And does it not now follow that, in this region, when I see a Plane and infer a Solid, I really see a Fourth unrecognized Dimension, not the same as colour, but existent, though infinitesimal and incapable of measurement?

And besides this, there is the Argument from Analogy of Figures.

Sphere. Analogy! Nonsense: what analogy?

I. Your Lordship tempts his servant to see whether he remembers the revelations imparted to him. Trifle not with me, my Lord; I crave, I thirst, for more knowledge. Doubtless we cannot *see* that other higher Spaceland now, because we have no eye in our stomachs. But, just as there *was* the realm of Flatland, though that poor puny Lineland Monarch could neither turn to left nor right to discern it, and just as there *was* close at hand, and touching my frame, the land of Three Dimensions, though I, blind senseless wretch, had no power to touch it, no eye in my interior to discern it, so of a surety there is a Fourth Dimension, which my Lord perceives with the inner eye of thought. And that it must exist my Lord himself has taught me. Or can he have forgotten what he himself imparted to his servant?

In One Dimension, did not a moving Point produce a Line with *two* terminal points?

In Two Dimensions, did not a moving Line produce a Square with *four* terminal points?

In Three Dimensions, did not a moving Square produce—did not this eye of mine behold it—that blessed Being, a Cube, with *eight* terminal points?

And in Four Dimensions shall not a moving Cube—alas, for

Analogy, and alas for the Progress of Truth, if it be not so—shall not, I say, the motion of a divine Cube result in a still more divine Organization with *sixteen* terminal points?

Behold the infallible confirmation of the Series, 2, 4, 8, 16: is not this a Geometrical Progression? Is not this—if I might quote my Lord's own words—'strictly according to Analogy'?

Again, was I not taught by my Lord that as in a Line there are *two* bounding Points, and in a Square there are *four* bounding Lines, so in a Cube there must be *six* bounding Squares? Behold once more the confirming Series, 2, 4, 6: is not this an Arithmetical Progression? And consequently does it not of necessity follow that the more divine offspring of the divine Cube in the Land of Four Dimensions, must have 8 bounding Cubes: and is not this also, as my Lord has taught me to believe, 'strictly according to Analogy'?

O, my Lord, my Lord, behold, I cast myself in faith upon conjecture, not knowing the facts; and I appeal to your Lordship to confirm or deny my logical anticipations. If I am wrong, I yield, and will no longer demand a Fourth Dimension; but, if I am right, my Lord will listen to reason.

I ask therefore, is it, or is it not, the fact, that ere now your countrymen also have witnessed the descent of Beings of a higher order than their own, entering closed rooms,* even as your Lordship entered mine, without the opening of doors or windows, and appearing and vanishing at will? On the reply to this question I am ready to stake everything. Deny it, and I am henceforth silent. Only vouchsafe an answer.

Sphere (*after a pause*). It is reported so. But men are divided in opinion as to the facts. And even granting the facts, they explain them in different ways. And in any case, however great may be the number of different explanations, no one has adopted or suggested the theory of a Fourth Dimension. Therefore, pray have done with this trifling, and let us return to business.

I. I was certain of it. I was certain that my anticipations would be fulfilled. And now have patience with me and answer me yet one more question, best of Teachers! Those who have thus appeared—no one knows whence—and have returned—no one knows whither—have they also contracted their sections and vanished somehow into that more Spacious Space, whither I now entreat you to conduct me?

Sphere (*moodily*). They have vanished, certainly—if they ever appeared. But most people say that these visions arose from the thought—you will not understand me—from the brain; from the perturbed angularity of the Seer.

I. Say they so? Oh, believe them not. Or if it indeed be so, that this other Space is really Thoughtland, then take me to that blessed Region where I in Thought shall see the insides of all solid things. There, before my ravished eye, a Cube, moving in some altogether new direction, but strictly according to Analogy, so as to make every particle of his interior pass through a new kind of Space with a wake of its own—shall create a still more perfect perfection than himself, with sixteen terminal Extra-solid angles, and Eight solid Cubes for his Perimeter. And once there, shall we stay our upward course? In that blessed region of Four Dimensions, shall we linger on the threshold of the Fifth, and not enter therein? Ah, no! Let us rather resolve that our ambition shall soar with our corporal ascent. Then, yielding to our intellectual onset, the gates of the Sixth Dimension shall fly open; after that a Seventh, and then an Eighth—

How long I should have continued I know not. In vain did the Sphere, in his voice of thunder, reiterate his commands of silence, and threaten me with the direst penalties if I persisted. Nothing could stem the flood of my ecstatic aspirations. Perhaps I was to blame; but indeed I was intoxicated with the recent draughts of Truth to which he himself had introduced me. However, the end was not long in coming. My words were cut short by a crash outside, and a simultaneous crash inside me, which impelled me through Space with a velocity that precluded speech. Down! down! down! I was rapidly descending; and I knew that return to Flatland was my doom. One glimpse, one last and never-to-be-forgotten glimpse I had of that dull level wilderness—which was now to become my Universe again—spread out before my eye. Then a darkness. Then a final, all-consummating thunder-peal; and, when I came to myself, I was once more a common creeping Square, in my Study at home, listening to the Peace-Cry of my approaching Wife.

20

How the Sphere encouraged me in a Vision

ALTHOUGH I had less than a minute for reflection, I felt, by a kind of instinct, that I must conceal my experiences from my Wife. Not that I apprehended, at the moment, any danger from her divulging my secret, but I know that to any Woman in Flatland the narrative of my adventures must needs be unintelligible. So I endeavoured to reassure her by some story, invented for the occasion, that I had accidentally fallen through the trap-door of the cellar, and had there lain stunned.

The Southward attraction in our country is so slight that even to a Woman my tale necessarily appeared extraordinary and well-nigh incredible; but my Wife, whose good sense far exceeds that of the average of her Sex, and who perceived that I was unusually excited, did not argue with me on the subject, but insisted that I was ill and required repose. I was glad of an excuse for retiring to my chamber to think quietly over what had happened. When I was at last by myself, a drowsy sensation fell on me; but before my eyes closed I endeavoured to reproduce the Third Dimension, and especially the process by which a Cube is constructed through the motion of a Square. It was not so clear as I could have wished; but I remembered that it must be 'Upward, and yet not Northward,' and I determined steadfastly to retain these words as the clue which, if firmly grasped, could not fail to guide me to the solution. So mechanically repeating, like a charm, the words, 'Upward yet not Northward,' I fell into a sound refreshing sleep.

During my slumber I had a dream. I thought I was once more by the side of the Sphere, whose lustrous hue betokened that he had exchanged his wrath against me for perfect placability. We were moving together towards a bright but infinitesimally small Point, to which my Master directed my attention. As we approached, methought there issued from it a slight humming noise as from one of your Spaceland blue-bottles,* only less resonant by far, so slight indeed that even in the perfect stillness of the Vacuum through which we soared, the sound reached not our ears till we checked our flight at a distance from it of something under twenty human diagonals.

'Look yonder,' said my Guide, 'in Flatland thou hast lived; of Lineland thou hast received a vision; thou hast soared with me to the heights of Spaceland; now, in order to complete the range of thy experience, I conduct thee downward to the lowest depth of existence, even to the realm of Pointland, the Abyss of No Dimensions.

'Behold yon miserable creature. That Point is a Being like ourselves, but confined to the non-dimensional Gulf. He is himself his own World, his own Universe; of any other than himself he can form no conception; he knows not Length, nor Breadth, nor Height, for he has had no experience of them; he has no cognizance even of the number Two; nor has he a thought of Plurality; for he is himself his One and All, being really Nothing. Yet mark his perfect self-contentment, and hence learn this lesson, that to be self-contented is to be vile and ignorant, and that to aspire is better than to be blindly and impotently happy. Now listen.'

He ceased; and there arose from the little buzzing creature a tiny, low, monotonous, but distinct tinkling, as from one of your Spaceland phonographs, from which I caught these words, 'Infinite beatitude of existence! It is; and there is none else beside It.'

'What,' said I, 'does the puny creature mean by "it"?' 'He means himself,' said the Sphere: 'have you not noticed before now, that babies and babyish people who cannot distinguish themselves from the world, speak of themselves in the Third Person? But hush!'

'It fills all Space,' continued the little soliloquizing Creature, 'and what It fills, It is. What It thinks, that It utters; and what It utters, that It hears; and It itself is Thinker, Utterer, Hearer, Thought, Word, Audition; it is the One, and yet the All in All. Ah, the happiness, ah, the happiness of Being!'

'Can you not startle the little thing out of its complacency?' said I. 'Tell it what it really is, as you told me; reveal to it the narrow limitations of Pointland, and lead it up to something higher.' 'That is no easy task,' said my Master; 'try you.'

Hereon, raising my voice to the uttermost, I addressed the Point as follows:

'Silence, silence, contemptible Creature. You call yourself the All in All, but you are the Nothing: your so-called Universe is a mere speck in a Line, and a Line is a mere shadow as compared with—' 'Hush, hush, you have said enough,' interrupted the Sphere, 'now listen, and mark the effect of your harangue on the King of Pointland.'

The lustre of the Monarch, who beamed more brightly than ever upon hearing my words, showed clearly that he retained his complacency; and I had hardly ceased when he took up his strain again. 'Ah, the joy, ah, the joy of Thought! What can It not achieve by thinking! Its own Thought coming to Itself, suggestive of Its disparagement, thereby to enhance Its happiness! Sweet rebellion stirred up to result in triumph! Ah, the divine creative power of the All in One! Ah, the joy, the joy of Being!'

'You see,' said my Teacher, 'how little your words have done. So far as the Monarch understands them at all, he accepts them as his own—for he cannot conceive of any other except himself—and plumes himself upon the variety of "Its Thought" as an instance of creative Power. Let us leave this God of Pointland to the ignorant fruition of his omnipresence and omniscience: nothing that you or I can do can rescue him from his self-satisfaction.'

After this, as we floated gently back to Flatland, I could hear the mild voice of my Companion pointing the moral of my vision, and stimulating me to aspire, and to teach others to aspire. He had been angered at first—he confessed—by my ambition to soar to Dimensions above the Third; but, since then, he had received fresh insight, and he was not too proud to acknowledge his error to a Pupil. Then he proceeded to initiate me into mysteries yet higher than those I had witnessed, showing me how to construct Extra-Solids by the motion of Solids, and Double Extra-Solids by the motion of Extra-Solids, and all 'strictly according to Analogy,' all by methods so simple, so easy, as to be patent even to the Female Sex.

21

How I tried to teach the Theory of Three Dimensions to my Grandson, and with what success

I AWOKE rejoicing, and began to reflect on the glorious career before me. I would go forth, methought, at once, and evangelize the whole of Flatland. Even to Women and Soldiers should the Gospel of Three Dimensions be proclaimed. I would begin with my Wife.

Just as I had decided on the plan of my operations, I heard the sound of many voices in the street commanding silence. Then followed a louder voice. It was a herald's proclamation. Listening attentively, I recognized the words of the Resolution of the Council, enjoining the arrest, imprisonment, or execution of any one who should pervert the minds of the people by delusions, and by professing to have received revelations from another World.

I reflected. This danger was not to be trifled with. It would be better to avoid it by omitting all mention of my Revelation, and by proceeding on the path of Demonstration—which after all, seemed so simple and so conclusive that nothing would be lost by discarding the former means. 'Upward, not Northward'—was the clue to the whole proof. It had seemed to me fairly clear before I fell asleep; and when I first awoke, fresh from my dream, it had appeared as patent as Arithmetic; but somehow it did not seem to me quite so obvious now. Though my Wife entered the room opportunely just at that moment, I decided, after we had interchanged a few words of commonplace conversation, not to begin with her.

My Pentagonal Sons were men of character and standing, and physicians of no mean reputation, but not great in mathematics, and, in that respect, unfit for my purpose. But it occurred to me that a young and docile Hexagon, with a mathematical turn, would be a most suitable pupil. Why therefore not make my first experiment with my little precocious Grandson, whose casual remarks on the meaning of 3^3 had met with the approval of the Sphere? Discussing the matter with him, a mere boy, I should be in perfect safety; for he would know nothing of the Proclamation of the Council; whereas I could not feel sure that my Sons—so greatly did their patriotism and reverence for the Circles predominate over mere blind affection—might not feel compelled to hand me over to the Prefect, if they found me seriously maintaining the seditious heresy of the Third Dimension.

But the first thing to be done was to satisfy in some way the curiosity of my Wife, who naturally wished to know something of the reasons for which the Circle had desired that mysterious interview, and of the means by which he had entered our house. Without entering into the details of the elaborate account I gave her,—an account, I fear, not quite so consistent with truth as my Readers in Spaceland might desire,—I must be content with saying that I succeeded at last in persuading her to return quietly to her household duties without eliciting from me any reference to the World of Three Dimensions. This done, I immediately sent for my Grandson; for, to confess the truth, I felt that all that I had seen and heard was in some strange way slipping away from me, like the image of a half-grasped, tantalizing dream, and I longed to essay my skill in making a first disciple.

When my Grandson entered the room I carefully secured the door. Then, sitting down by his side and taking our mathematical tablets—or, as you would call them, Lines—I told him we would resume the lesson of yesterday. I taught him once more how a Point by motion in One Dimension produces a Line, and how a straight Line in Two Dimensions produces a Square. After this, forcing a laugh, I said, 'And now, you scamp, you wanted to make me believe that a Square may in the same way by motion "Upward, not Northward," produce another figure, a sort of extra Square in Three Dimensions. Say that again, you young rascal.'

At this moment we heard once more the herald's 'O yes! O yes!'* outside in the street proclaiming the Resolution of the Council. Young though he was, my Grandson—who was unusually intelligent for his age, and bred up in perfect reverence for the authority of the Circles—took in the situation with an acuteness for which I was quite unprepared. He remained silent till the last words of the Proclamation had died away, and then, bursting into tears, 'Dear Grandpapa,' he said, 'that was only my fun, and of course I meant nothing at all by it; and we did not know anything then about the new Law; and I don't think I said anything about the Third Dimension; and I am sure I did not say one word about "Upward, not Northward," for that would be such nonsense, you know. How could a thing move Upward, and not Northward? Upward, and not Northward! Even if I were a baby, I could not be so absurd as that. How silly it is! Ha! ha! ha!'

'Not at all silly,' said I, losing my temper; 'here for example, I take this Square,'—and, at the word, I grasped a moveable Square, which was lying at hand—'and I move it, you see, not Northward but—yes, I move it Upward—that is to say, not Northward, but I move it somewhere—not exactly like this, but somehow—' Here I brought my sentence to an inane conclusion, shaking the Square about in a purposeless manner, much to the amusement of my Grandson, who burst out laughing louder than ever, and declared that I was not teaching him, but joking with him. So saying he unlocked the door and ran out of the room; and thus ended my first attempt to convert a pupil to the Gospel of Three Dimensions.

22

How I then tried to diffuse the Theory of Three Dimensions by other means, and of the result

MY failure with my Grandson did not encourage me to communicate my secret to others of my household; yet neither was I led by it to despair of success. Only I saw that I must not wholly rely on the catch-phrase 'Upward, not Northward,' but must rather endeavour to seek a demonstration by setting before the public a clear view of the whole subject; and for this purpose it seemed necessary to resort to writing.

So I devoted several months in privacy to the composition of a treatise on the mysteries of Three Dimensions. Only, with the view of evading the Law, if possible, I spoke not of a physical Dimension, but of a Thoughtland whence, in theory, a Figure could look down upon Flatland and see simultaneously the insides of all things, and where it was possible that there might be supposed to exist a Figure environed, as it were, with six Squares, and containing eight terminal Points. But in writing this book I found myself sadly hampered by the impossibility of drawing such diagrams as were necessary for my purpose; for of course, in our country of Flatland, there are no tablets but Lines, and no diagrams but Lines, all in one straight Line and only distinguishable by difference of size and brightness; so that, when I had finished my treatise (which I entitled 'Through Flatland to Thoughtland'*) I could not feel certain that many would understand my meaning.

Meanwhile my life was under a cloud. All pleasures palled upon me; all sights tantalized and tempted me to outspoken treason, because I could not but compare what I saw in Two Dimensions with what it really was if seen in Three, and could hardly refrain from making my comparisons aloud. I neglected my clients and my own business to give myself to the contemplation of the mysteries which I had once beheld, yet which I could impart to no one, and found daily more difficult to reproduce even before my own mental vision.

One day, about eleven months after my return from Spaceland, I tried to see a Cube with my eye closed, but failed; and though I succeeded afterwards, I was not then quite certain (nor have I been ever afterwards) that I had exactly realized the original. This made me more melancholy than before, and determined me to take some step; yet what, I knew not. I felt that I would have been willing to sacrifice my life for the Cause, if thereby I could have produced conviction. But if I could not convince my Grandson, how could I convince the highest and most developed Circles in the land?

And yet at times my spirit was too strong for me, and I gave vent to dangerous utterances. Already I was considered heterodox if not treasonable, and I was keenly alive to the dangers of my position; nevertheless I could not at times refrain from bursting out into suspicious or half-seditious utterances, even among the highest Polygonal and Circular society. When, for example, the question arose about the treatment of those lunatics who said that they had received the power of seeing the insides of things, I would quote the saying of an ancient Circle, who declared that prophets and inspired people are always considered by the majority to be mad; and I could not help occasionally dropping such expressions as 'the eye that discerns the interiors of things,' and 'the all-seeing land:' once or twice I even let fall the forbidden terms 'the Third and Fourth Dimensions.' At last, to complete a series of minor indiscretions, at a meeting of our Local Speculative Society held at the palace of the Prefect himself,—some extremely silly person having read an elaborate paper exhibiting the precise reasons why Providence has limited the number of Dimensions to Two, and why the attribute of omnividence is assigned to the Supreme alone—I so far forgot myself as to give an exact account of the whole of my voyage with the Sphere into Space, and to the Assembly Hall in our Metropolis, and then to Space again, and of my return home, and of everything that I had seen and heard in fact or vision. At first, indeed, I pretended that

I was describing the imaginary experiences of a fictitious person; but my enthusiasm soon forced me to throw off all disguise, and finally, in a fervent peroration, I exhorted all my hearers to divest themselves of prejudice and to become believers in the Third Dimension.

Need I say that I was at once arrested and taken before the Council?

Next morning, standing in the very place where but a very few months ago the Sphere had stood in my company, I was allowed to begin and to continue my narration unquestioned and uninterrupted. But from the first I foresaw my fate; for the President, noting that a guard of the better sort of Policemen was in attendance, of angularity little, if at all, under 55°, ordered them to be relieved before I began my defence, by an inferior class of 2° or 3°. I knew only too well what that meant. I was to be executed or imprisoned, and my story was to be kept secret from the world by the simultaneous destruction of the officials who had heard it; and, this being the case, the President desired to substitute the cheaper for the more expensive victims.

After I had concluded my defence, the President, perhaps perceiving that some of the junior Circles had been moved by my evident earnestness, asked me two questions:—

1. Whether I could indicate the direction which I meant when I used the words 'Upward, not Northward'?

2. Whether I could by any diagrams or descriptions (other than the enumeration of imaginary sides and angles) indicate the Figure I was pleased to call a Cube?

I declared that I could say nothing more, and that I must commit myself to the Truth, whose cause would surely prevail in the end.

The President replied that he quite concurred in my sentiment, and that I could not do better. I must be sentenced to perpetual imprisonment; but if the Truth intended that I should emerge from prison and evangelize the world, the Truth might be trusted to bring that result to pass. Meanwhile I should be subjected to no discomfort that was not necessary to preclude escape, and, unless I forfeited the privilege by misconduct, I should be occasionally permitted to see my brother, who had preceded me to my prison.

Seven years have elapsed and I am still a prisoner, and—if I except the occasional visits of my brother—debarred from all companionship save that of my jailers. My brother is one of the best of Squares, just, sensible, cheerful, and not without fraternal affection; yet I must confess that my weekly interviews, at least in one respect, cause me the bitterest pain. He was present when the Sphere manifested himself in the Council Chamber; he saw the Sphere's changing sections; he heard the explanation of the phenomena then given to the Circles. Since that time, scarcely a week has passed during seven whole years, without his hearing from me a repetition of the part I played in that manifestation, together with ample descriptions of all the phenomena in Spaceland, and the arguments for the existence of Solid things derivable from Analogy. Yet—I take shame to be forced to confess it—my brother has not yet grasped the nature of the Third Dimension, and frankly avows his disbelief in the existence of a Sphere.

Hence I am absolutely destitute of converts, and, for aught that I can see, the millennial Revelation has been made to me for nothing. Prometheus up in Spaceland was bound for bringing down fire for mortals, but I—poor Flatland Prometheus—lie here in prison for bringing down nothing to my countrymen. Yet I exist in the hope that these memoirs, in some manner, I know not how, may find their way to the minds of humanity in Some Dimension, and may stir up a race of rebels who shall refuse to be confined to limited Dimensionality.

That is the hope of my brighter moments. Alas, it is not always so. Heavily weighs on me at times the burdensome reflection that I cannot honestly say I am confident as to the exact shape of the once-seen, oft-regretted Cube; and in my nightly visions the mysterious precept, 'Upward, not Northward,' haunts me like a soul-devouring Sphinx. It is part of the martyrdom which I endure for the cause of the Truth that there are seasons of mental weakness, when Cubes and Spheres flit away into the background of scarce-possible existence; when the Land of

Three Dimensions seems almost as visionary as the Land of One or None; nay, when even this hard wall that bars me from my freedom, these very tablets on which I am writing, and all the substantial realities of Flatland itself, appear no better than the offspring of a diseased imagination, or the baseless fabric of a dream.

EXPLANATORY NOTES

1 *facsimile title page from Blackwell third edition*: quotation at the top of the page, 'O day and night, but this is wondrous strange' (*Hamlet*, I. v. 164), is Horatio's perplexed response upon hearing the ghost of Hamlet's father, who seconds Hamlet's demand that Horatio and Marcellus swear not to reveal what they have seen. This is the first of several quotations from Shakespeare's plays. Abbott was well versed in the subject, having published *A Shakespearean Grammar* in 1869. His advanced students at the City of London School studied Shakespearian plays and performed scenes from them each year. In this context 'A Romance' refers to a work of fiction that deals with characters, events, and/or places remote from the circumstances of ordinary life. The quotation at the foot of the page, 'And therefore as a stranger give it welcome' (*Hamlet* I. v. 165), is Hamlet's more open-minded response to Horatio concerning the ghost of his father.

3 *title page*: although there is evidence that some contemporaries knew that Abbott was the author of *Flatland* (for instance, see the reviews in *Oxford Magazine, Literary World*, and *City of London School Magazine*), Abbott publicly identified himself as the author of *Flatland* on the title page and again on p. 29 of *The Spirit on the Waters: The Evolution of the Divine from the Human* (1897). 'Fie, fie, how franticly I square my talk!' (*Titus Andronicus*, III. ii. 31) is Titus's impatient comment to his brother Marcus, who has just reminded him of Titus's own mutilation and that of his daughter Lavinia by their enemies. 'Square' in this context means to regulate or to adjust.

5 *To . . . H.C. in Particular*: Howard Candler, fellow mathematician and close friend of Abbott's. Abbott identifies him in *Apologia* (1907), p. xiii.

7 *the editor*: the anonymous editor speaks for Abbott himself, who uses the Preface to reply to criticism made by readers of the first edition.

9 *Extra-Cubes*: hypercubes or tesseracts, the four-dimensional projection of a cube.

'One touch of Nature makes all worlds akin': after Shakespeare, *Troilus and Cressida*, III. iii. 174, 'One touch of nature makes the whole world kin.' The 'One touch of nature' that Ulysses refers to in this speech is the common human weakness of neglecting traditional things of value in favour of whatever is currently fashionable.

13 *'Be patient, for the world is broad and wide'*: Shakespeare, *Romeo and Juliet*, III. iii. 16, the friar's words of counsel to Romeo, who has just learned that he has been banished from Verona.

18 *a point of breeding . . . always to give her the North side of the way*: the parallel in traditional Spaceland etiquette is that a man accompanying a

lady is supposed to take the outside of the pavement to protect her from being splashed by passing traffic.

21 *Our Professional Men and Gentlemen*: these groups were higher in social distinction than the merely 'middle-class' merchants or Equilaterals. The three traditional professions were medicine, the Church, and the law (to which the Square belongs), although by Abbott's era other types of white-collar work were included in this category. The definition of the gentleman was a focus of much Victorian debate. Traditionally identified by leisure, birth, and wealth, the gentleman was redefined over the course of the nineteenth century to emphasize moral and intellectual traits over strictly economic ones. By the 1880s, having received the kind of liberal education and character training offered by a public school provided a widely accepted evidence of gentlemanly status (see also the note for p. 40).

It is a Law of Nature . . .: the remainder of this chapter outlines a system of social mobility with various resemblances to Abbott's own society. See the section on 'The Shape of Society in Flatland' in the Introduction for more details.

25 *Invisible Cap*: Perseus in Greek legend is able to slay the Medusa by wearing a cap that renders him invisible. Similar caps are featured in European and Punjabi folk tales.

26 *Peace-cry*: a humming sound (see p. 29) that women are required to make to announce their presence; presumably the noise helps keep the peace by protecting men from women's sharp points.

St. Vitus's Dance: a convulsive disorder causing involuntary motions in the limbs and face.

Teachers of Board Schools: the fact that Board Schools, established after the Education Act of 1870, educated children at state expense made them less prestigious than schools supported by parental fees. Board School teachers, although drawn largely from the same classes as their pupils—the working and lower middle—felt that their educational achievements earned them the right to be considered gentlemen and ladies. Such aspirations were often resented by their social betters, like the parish rector of South Wytham quoted in the article cited here, who snubs the local schoolmaster by informing him that in the future he should enter by the rectory's kitchen door like other social inferiors. Abbott no doubt agreed with *The Spectator*'s opinion that such clerical snobbery weakened society's regard for the Church of England.

27 *well-modulated undulation of the back in our ladies of Circular rank*: proper upright posture and graceful movement were part of the deportment taught in girls' finishing schools and considered an essential indication of good breeding among the upper classes.

28 *seasonable simulations*: that is, the kind of dissembling rhetoric that Flatland males normally use to placate females; compare, for instance, the deference that conceals contempt described near the end of Chapter 12.

29 *the very Laws of Evolution*: Charles Darwin had introduced his theory of evolution in *The Origin of Species* (1859). He extended its implications to human beings in *The Descent of Man* (1872), which included an evolutionary explanation for the alleged inferiority of women to men. For a discussion of the influence of evolution and other scientific theories on conceptions of class and gender in Flatland, see 'Justifying the Status Quo' and 'Female Flatlanders' in the Introduction.

32 *University of Wentbridge*: a pun on the University of Cambridge, perhaps with the implication that teaching there is behind the times, at least in comparison with the more modern and practical studies that Abbott himself supported at the City of London School.

35 *a reaction in favour of 'the cheap system'*: in addition to criticizing callous attitudes toward the poor, Abbott may here be satirizing attitudes toward state-supported education. Many members of the upper classes were concerned that poor children should be educated as cheaply as possible. As Robert Lowe, the government official in charge of popular education in the 1860s, put it, 'if education is not cheap, it should be efficient; if it is not efficient, it should be cheap'. To ensure that the state was getting its money's worth, Lowe instituted a policy called 'payment by results' in 1862. It linked the amount of government support a school received to the number of students who passed standardized tests in basic subjects each year. Widely unpopular with teachers for the way it narrowed the scope of education, 'payment by results' remained in effect in Board Schools until the early 1880s.

40 *Rustication*: a form of suspension from academic studies.

spend a third of his life in abstract studies: this section suggests the contrast between the elementary and practical skills provided for working- and lower-middle-class children destined for early entry into the workforce and the classical studies that formed the basis of liberal education at 'higher Seminaries of an exclusive character': privately supported schools like traditional British public schools (so called because their enrolment was not necessarily restricted to students from a particular area). The non-utilitarian (or 'abstract') character of classical education ensured its value as a status symbol; only those with sufficient leisure and wealth could afford to educate their sons in this way. The exclusive social function of Latin and Greek was reinforced by the fact that both were necessary for university admission. However, as Abbott goes on to imply here, the investment in this prestigious kind of liberal education usually paid off by ensuring that graduates could move with ease and security in upper-class society. Indeed, by the 1880s, a public school education (and the kind of character training that was assumed to accompany its classical curriculum) had become the de facto criterion for laying claim to the status of a gentleman.

42 *Of Irregular Figures*: in this chapter Abbott satirizes Victorians' intolerance

for non-conformity; see 'Justifying the Status Quo' in the Introduction for further discussion.

45 *Via Media*: a 'middle way' or compromise position.

46 *Chromatistes*: from *chrōma*, Greek for colour.

48 *the very language and vocabulary of the period*: in some respects the discovery of painting in Flatland suggests the flowering of humanism in the English Renaissance and the richness of language in authors like Shakespeare, in other respects the departure from neoclassical restraint in the work of the British Romantic poets.

49 *Of the Universal Colour Bill*: The radical push for equality described in Chapter 9 suggests similarities to various democratic movements: peasant rebellions in the Middle Ages, the French Revolution, the push for political rights by working-class men in England during the Chartist movement in the 1840s (which included 'universal manhood suffrage' among its demands), the agitation for women's rights, which gained strength in the second half of the century, as well as the more successful movements to widen the franchise in the three Reform Bills (1832, 1867, and 1884).

54 *Pantocyclus*: from the Greek *pantos* meaning all or every and Latin *cyclus* meaning circle.

59 *the race shall become less fertile*: this suggests late Victorian fears that the upper classes were being out-reproduced by the classes below them (see the discussion of eugenics in 'Justifying the Status Quo' in the Introduction).

61 *the vain belief that conduct depends upon will, effort, training, encouragement, praise, or anything else but Configuration*: the evolutionary and eugenic stress on heredity as the determinant of behaviour opposed popular Victorian faith in the power of self-help and will to improve oneself. As seen in the ensuing discussion about the Square's grandsons, this strict adherence to what we might call biological determinism threatened to undermine free will and moral responsibility and was tempered in practice by the assumption that individuals could be educated to choose between right and wrong. See 'Justifying the Status Quo' in the Introduction for further discussion of the respective claims of nature and nurture in shaping individual behaviour during Abbott's era.

63 *no blot on her escutcheon*: no stain on her reputation (an escutcheon is the shield upon which a family's coat of arms would be depicted, hence figuratively representing its reputation or good name).

'an excellent thing in Woman': after Shakespeare, *King Lear*, v. iii. 273–4, where Lear eulogizes his daughter Cordelia by saying, 'Her voice was ever soft, | Gentle, and low—an excellent thing in woman.'

64 *since Women are deficient in Reason but abundant in Emotion*: this sums up views widely held by Victorian scientists and laymen. Such views also

reinforced the belief that middle-class men and women should occupy 'separate spheres', women in charge of home and family, men responsible for business and governance. See 'Female Flatlanders' in the Introduction for a discussion of Abbott's treatment of gender issues.

67 *'O brave new worlds, | That have such people in them!'*: after Shakespeare, *The Tempest*, V. i. 183–4, where the sheltered Miranda exclaims, 'O brave new world | That has such people in't' upon meeting other human beings in addition to Ferdinand, the son of the Duke of Naples. Her father Prospero replies, ' 'Tis new to thee.'

69 *Long Vacation*: in Spaceland, the summer vacation at Oxford and Cambridge Universities, which is considerably longer than the recesses at Christmas and Easter.

76 *lilliputian grasshoppers*: tiny grasshoppers, such as would be found in the miniaturized world that Gulliver encounters on the island of Lilliput in the first part of *Gulliver's Travels*.

83 *second Millennium*: after 1999, the third millennium would begin. This error was corrected in the third edition of *Flatland*.

86 *Stranger. (To himself). What must I do?*: this marks the beginning of one of the additions to the new and revised edition mentioned in the footnote on p. 9 of the Preface. The addition extends to '*Stranger*. (*To himself*). I can do neither' on p. 87.

89 *juggler*: a trickster or conjurer.

102 *a second Prometheus*: Prometheus was a Titan punished by Zeus for giving fire to humankind; he was chained to a rock, where his liver would regenerate each night after being eaten by eagles each day.

103 *I. Not inconceivable, my Lord, to me*: this marks the beginning of the second addition referred to in the Preface (p. 9 n. 1), which ends with the Sphere's comment 'Analogy! Nonsense: what analogy?'

105 *entering closed rooms*: this suggests biblical miracles and spiritualist manifestations, which some nineteenth-century believers attributed to the fourth dimension. For Abbott's own views, see the section on 'Flatland and Nineteenth-Century Geometries' in the Introduction.

109 *blue-bottles*: bluebottle flies, so called because of the colour of their bodies.

112 *'O yes! O yes!'*: from 'oyez', the imperative form of the French verb *ouïr*, to hear; hence 'hear ye!'

114 *'Through Flatland to Thoughtland'*: compare the title of Abbott's 1877 theological treatise *Through Nature to Christ, or, The Ascent of Worship Through Illusion to the Truth*.

118 *final illustration*: 'The baseless fabric of this vision melted into air, into thin air | Such stuff as dreams made of', is after Shakespeare (*The Tempest*, IV. i. 148–57), where Prospero reveals that what Ferdinand and others have witnessed were visions created by himself:

Our revels now are ended. These our actors,
As I foretold you, were all spirits and
Are melted into air, into thin air;
And, like the baseless fabric of this vision,
The cloud-capped tow'rs, the gorgeous palaces,
The solemn temples, the great globe itself,
Yea, all which it inherit, shall dissolve,
And, like this insubstantial pageant faded,
Leave not a rack behind. We are such stuff
As dreams are made on, and our little life
Is rounded with a sleep.

MORE ABOUT **OXFORD WORLD'S CLASSICS**

American Literature

British and Irish Literature

Children's Literature

Classics and Ancient Literature

Colonial Literature

Eastern Literature

European Literature

Gothic Literature

History

Medieval Literature

Oxford English Drama

Poetry

Philosophy

Politics

Religion

The Oxford Shakespeare

A SELECTION OF **OXFORD WORLD'S CLASSICS**

	Late Victorian Gothic Tales
Jane Austen	**Emma** **Mansfield Park** **Persuasion** **Pride and Prejudice** **Selected Letters** **Sense and Sensibility**
Mrs Beeton	**Book of Household Management**
Mary Elizabeth Braddon	**Lady Audley's Secret**
Anne Brontë	**The Tenant of Wildfell Hall**
Charlotte Brontë	**Jane Eyre** **Shirley** **Villette**
Emily Brontë	**Wuthering Heights**
Robert Browning	**The Major Works**
John Clare	**The Major Works**
Samuel Taylor Coleridge	**The Major Works**
Wilkie Collins	**The Moonstone** **No Name** **The Woman in White**
Charles Darwin	**The Origin of Species**
Thomas De Quincey	**The Confessions of an English Opium-Eater** **On Murder**
Charles Dickens	**The Adventures of Oliver Twist** **Barnaby Rudge** **Bleak House** **David Copperfield** **Great Expectations** **Nicholas Nickleby** **The Old Curiosity Shop** **Our Mutual Friend** **The Pickwick Papers**

A SELECTION OF OXFORD WORLD'S CLASSICS

Charles Dickens	A Tale of Two Cities
George du Maurier	Trilby
Maria Edgeworth	Castle Rackrent
George Eliot	Daniel Deronda The Lifted Veil and Brother Jacob Middlemarch The Mill on the Floss Silas Marner
Susan Ferrier	Marriage
Elizabeth Gaskell	Cranford The Life of Charlotte Brontë Mary Barton North and South Wives and Daughters
George Gissing	New Grub Street The Odd Women
Edmund Gosse	Father and Son
Thomas Hardy	Far from the Madding Crowd Jude the Obscure The Mayor of Casterbridge The Return of the Native Tess of the d'Urbervilles The Woodlanders
William Hazlitt	Selected Writings
James Hogg	The Private Memoirs and Confessions of a Justified Sinner
John Keats	The Major Works Selected Letters
Charles Maturin	Melmoth the Wanderer
John Ruskin	Selected Writings
Walter Scott	The Antiquary Ivanhoe

A SELECTION OF **OXFORD WORLD'S CLASSICS**

WALTER SCOTT	**Rob Roy**
MARY SHELLEY	**Frankenstein** **The Last Man**
ROBERT LOUIS STEVENSON	**Strange Case of Dr Jekyll and Mr Hyde and Other Tales** **Treasure Island**
BRAM STOKER	**Dracula**
JOHN SUTHERLAND	**So You Think You Know Jane Austen?** **So You Think You Know Thomas Hardy?**
WILLIAM MAKEPEACE THACKERAY	**Vanity Fair**
OSCAR WILDE	**The Importance of Being Earnest and Other Plays** **The Major Works** **The Picture of Dorian Gray**
ELLEN WOOD	**East Lynne**
DOROTHY WORDSWORTH	**The Grasmere and Alfoxden Journals**
WILLIAM WORDSWORTH	**The Major Works**

TROLLOPE IN OXFORD WORLD'S CLASSICS

ANTHONY TROLLOPE

The American Senator
An Autobiography
Barchester Towers
Can You Forgive Her?
The Claverings
Cousin Henry
The Duke's Children
The Eustace Diamonds
Framley Parsonage
He Knew He Was Right
Lady Anna
Orley Farm
Phineas Finn
Phineas Redux
The Prime Minister
Rachel Ray
The Small House at Allington
The Warden
The Way We Live Now

A SELECTION OF OXFORD WORLD'S CLASSICS

THOMAS AQUINAS	**Selected Philosophical Writings**
FRANCIS BACON	**The Essays**
WALTER BAGEHOT	**The English Constitution**
GEORGE BERKELEY	**Principles of Human Knowledge** and **Three Dialogues**
EDMUND BURKE	**A Philosophical Enquiry into the Origin of Our Ideas of the Sublime and Beautiful** **Reflections on the Revolution in France**
CONFUCIUS	**The Analects**
DESCARTES	**A Discourse on the Method**
ÉMILE DURKHEIM	**The Elementary Forms of Religious Life**
FRIEDRICH ENGELS	**The Condition of the Working Class in England**
JAMES GEORGE FRAZER	**The Golden Bough**
SIGMUND FREUD	**The Interpretation of Dreams**
THOMAS HOBBES	**Human Nature** and **De Corpore Politico** **Leviathan**
DAVID HUME	**Selected Essays**
NICCOLÒ MACHIAVELLI	**The Prince**
THOMAS MALTHUS	**An Essay on the Principle of Population**
KARL MARX	**Capital** **The Communist Manifesto**
J. S. MILL	**On Liberty and Other Essays** **Principles of Political Economy** and **Chapters on Socialism**
FRIEDRICH NIETZSCHE	**Beyond Good and Evil** **The Birth of Tragedy** **On the Genealogy of Morals** **Thus Spoke Zarathustra** **Twilight of the Idols**

A SELECTION OF **OXFORD WORLD'S CLASSICS**
